Monoclonal Antibodies in Tumor Therapy

Contributions to Oncology
Beiträge zur Onkologie

Vol. 32

Series Editors
S. Eckhardt, Budapest; *J. H. Holzner,* Wien;
G. A. Nagel, Göttingen

Basel · München · Paris · London · New York · New Delhi · Singapore · Tokyo · Sydney

Monoclonal Antibodies in Tumor Therapy

Present Stage, Chances and Limitations

*H.H. Sedlacek, G. Schulz, A. Steinstraesser, L. Kuhlmann,
A. Schwarz, L. Seidel, G. Seemann, H.-P. Kraemer, K. Bosslet,*
Marburg and Frankfurt/Griesheim

18 figures and 57 tables, 1988

RC254
B44
v.32
1988

Basel · München · Paris · London · New York · New Delhi · Singapore · Tokyo · Sydney

Contributions to Oncology
Beiträge zur Onkologie

Drug Dosage
The authors and the publisher have exerted every effort to ensure that drug selection and dosage set forth in this text are in accord with current recommendations and practice at the time of publication. However, in view of ongoing research, changes in government regulations, and the constant flow of information relating to drug therapy and drug reactions, the reader is urged to check the package insert for each drug for any change in indications and dosage and for added warnings and precautions. This is particularly important when the recommended agent is a new and/or infrequently employed drug.

All rights reserved.
No part of this publication may be translated into other languages, reproduced or utilized in any form or by any means, electronic or mechanical, including photocopying, recording, microcopying, or by any information storage and retrieval system, without permission in writing from the publisher.

© Copyright 1988 by S. Karger GmbH, P.O. Box 1724, D-8034 Germering/München and
S. Karger AG, Postfach, CH-4009 Basel
Printed in Germany by Lipp GmbH, München
ISBN 3-8055-4763-3

Contents

Foreword	VI
Preface	VIII
Abbreviations	IX
Introduction	1
Tumor Antigens	5
Generation, Selection, Properties and Quality Control of Murine Monoclonal Antibodies	20
Generation and Selection	20
Affinity	25
Fc-Mediated Functions	28
Quality Control	31
Radioimmunolocalization	38
Selection of Radionuclides	39
Methods of Labelling Antibodies with Radionuclides	42
Analysis of Immunoreactivity	49
Experimental Tumor Imaging	53
Clinical Data	59
Specific Radioimmunotherapy	68
Physicochemical and Biological Considerations	68
Clinical Trials	76
Specific Chemoimmunotherapy	81
Conjugation of Cytostatics or Toxins	81
Analysis and Parameters of Antitumoral Activity	85
Clinical Data and Pharmacokinetic Considerations	94
Specific Immunotherapy	101
Mechanisms of Tumor Cell Destruction	101
Treatment of Patients with Leukemia and Lymphoma	104
Treatment of Patients with Solid Tumors	106
Immune Response against Murine Immunoglobulins	109
Bone Marrow Purging of Neoplastic Cells	114
Human or Humanized Monoclonal Antibodies	119
Human Monoclonal Antibodies	119
Chimeric Antibodies and Humanization	121
Conclusion	127
Acknowledgement	132
References	133

Foreword

This volume is dedicated to Professor Dr. phil. Hans Gerhard Schwick on the occasion of his sixtieth birthday. Its theme, 'Monoclonal Antibodies in Tumor Therapy', is an attempt to review some of the most important achievements which have emerged during the rapid progress of a science to which H. G. Schwick has devoted his life's work.

That the company, founded by Emil von Behring at the turn of the century, should be under his leadership is without doubt a result, not only of his creative flair for science, but also of his talent for enthusing his colleagues with the spirit of cooperation and for demonstrating the vital importance of his science.

'His science' — the sphere of immunology and the related fields of biochemistry, hemostaseology and clinical chemistry. This is the home of his investigations on pepsin cleavage of human immunoglobulins, discoveries which made possible wide-ranging anti-infectious intravenous plasma therapy. This is also the field that has nurtured developments in plasma fractionation and enabled the production of highly purified factor concentrates, inhibitors and trace proteins, a field to whose advances his work has contributed so much.

The path of his career led from the laboratory bench, under the guidance of his teacher and mentor H. E. Schultze, to cooperation with a whole range of research institutes and universities. His active membership and participation in committees of numerous scientific organizations, and the various distinctions awarded him, attest to the high degree of international recognition he has received. This, coupled with his successes in the field of pharmaceutical research, was not without consequence: H. G. Schwick joined the Board of Directors at Behringwerke in 1968, and has now been Chairman since 1980.

In the tradition founded by Emil von Behring, H. G. Schwick possesses the ability to demonstrate the practical value behind this science of the immunological processes in the human body. And in medicine today, the trend is clearly towards establishing the immunological determinants of every disease so that knowledge can be gained of the conditions underlying their pathogenesis.

H. G. Schwick has always viewed immunology as a practical science, and his many experiments have provided clear and concise illustrations of complex regulatory processes. Perhaps it is precisely this eye for the essential which has helped pave his way to success.

Together with all his friends and co-workers, we wish him continued success along this path, and all the best for his sixtieth birthday.

Marburg, April 1988

W. Bernhardt
E.-G. Afting

Preface

Monoclonal antibodies are under discussion as representing a new chance in tumor immunotherapy. Indeed, the localization of tumors in patients with the use of radiolabelled monoclonal antibodies is already possible. Provided all the conditions for standardizing this technique are fulfilled and all possibilities now available to optimize immunoscintigraphy are taken advantage of, immunoscintigraphy of tumors is now ready for clinical routine use. This may provide hope in tumor immunotherapy. However, the dosimetric studies clearly revealed that the amount of antibody localizing at the tumor site compared to normal tissue is too low to reach tumor-specific cytotoxicity without intolerable side effects by radionuclides, toxins or cytostatics linked to the antibody.

Future research has thus to concentrate on this problem, i.e. to look for new antibody specificities with enhanced tumor localization and to develop new antibody preparations with reduced binding to normal tissue. Tumor immunotherapy with conjugated and unconjugated monoclonal antibodies suffers from the xenogeneity of the murine immunoglobulin. Consequently, the chance should be taken to humanize these antibodies and to evaluate the potency of these preparations in tumor therapy. In case the anti-idiotypic response of the patient against the applied murine immunoglobulin proves to be essential for the effect in tumor therapy, a new way of treating tumor patients by immunization would be opened up.

All in all, we are just beginning to evaluate the potency of monoclonal antibodies in tumor therapy. There are a number of reasons to be optimistic, but they should not tempt us to overlook the problems which have to be solved.

Marburg, April 1988 *H. H. Sedlacek*

Abbreviations

ADCC	antibody-dependent cytotoxicity
AML	acute myeloid leukemia
ANLL	acute non-lymphocytic leukemia
C	constant region of IgG
CDR	complementary determining region
CLL	chronic lymphatic leukemia
CTCL	chronic T cell leukemia
CMC	complement-mediated cytolysis
DTPA	diethylene-triamine-penta-acetic acid
EBV	Epstein Barr virus
ECT	emission computer tomography
EGF	epithelial growth factor
ELISA	enzyme-linked immunosorbent assay
Fab	Fab fragment of IgG
F(ab')$_2$	F(ab')$_2$ fragment of IgG
Fc	fragment crystalline of IgG
GM-CSF	granulocyte macrophages colony-stimulating factor
HAMA	human antibodies against murine antibodies
HMFG	high molecular milk fat globulin
Il-2	Interleukin 2
LCL	lymphatic cell leukemia
MAb	monoclonal antibody
MAP	mouse antibody production test
MW	molecular weight
NCA	non-specific cross-reacting antigen
NK cells	natural killer cells
PDGF	platelet-derived growth factor
RES	reticuloendothelial system
SCLC	small cell lung carcinoma
TAA	tumor-associated antigen
TAE	tumor-associated epitopes
TRF	transferrin
TuMAb	tumor monoclonal antibodies
V	variable region of IgG

Introduction

Based on the assumption that tumors may induce an immune response, i.e. antibody response [176], the 'immune surveillance' hypothesis [102, 103, 687] claims that aberrant and malignant cells can continuously be recognized and eliminated by the host's own immune system.

A prerequisite for the immune surveillance is the existence of tumor-associated or even tumor-specific antigens which enable the immune system to recognize tumor cells as being foreign and to develop a tumor-cell-specific cellular and/or humoral immune response. Any insufficiency of the immune system consequently allows aberrant cells to grow and to develop, for instance, into a clinically apparent tumor disease. According to the immune surveillance hypothesis, a causative treatment of tumor diseases would be the application of cytotoxic cells and/or antibodies specific to the tumor (for review, see [604]).

Thus for a long time attempts were made to treat tumor diseases using antisera and antibodies isolated either from animals immunized with the respective tumor or even from patients in tumor remission (see table I). This kind of treatment, however, only showed limited success. The reasons for failure were manifold: the respective tumor target cells, the specificity and quality of the antibody used, as well as the actual general condition of the host, and the local anatomical situation at the tumor site (see table II).

The failure of tumor therapy using antibodies, together with more or less similar failures of other kinds of tumor immunotherapy, cast doubts upon the validity of the immune surveillance hypothesis. These doubts have been augmented by the difficulty and even impossibility of proving the existence of tumor-specific antigens in spontaneous autochthonous tumors.

The use of antibodies in tumor therapy thus turned out to be a blind alley which, however, surprisingly opened into new research avenues through the development of hybridoma technology [370] for the production of monoclonal antibodies. Today this technique has been so greatly improved upon that no basic hindrance exists to produce unlimited amounts of at least murine antibodies with selected homogeneous specificity.

Table I. Immunotherapy of tumor patients with antibody preparations

Year	Author	Preparation	Source	Specificity
1895	Hericourt and Richet	antisera	dogs; monkeys	immunized with respective tumor
1901	Boeri	antisera	goat	immunized with respective tumor
1958	Murray	IgG	horses	immunized with respective tumor
1959	Buinauskas et al.	IgG	sheep	immunized with respective tumor
1960	Sumner and Foraker	whole blood	man	patients with same type of tumor in regression
1968	Laszlo et al.	sera	man	isoantibodies against lymphocytes (CLL)

Table II. Reasons for the limited success of antibodies in tumor therapy

Target cell
- Lack of any tumor-specific antigen
- Low degree of expression of tumor antigens
    - low amount per cell
    - steric hindrance
    - masking of antigens
- Qualitative and quantitative variation in the expression of antigens
- Release (shedding) of antigens and peripheral neutralization of antitumoral antibodies

Antibodies
- Insufficient specificity, affinity and quantity
- Insufficient potency to induce effector mechanisms for cytolysis (complement activation; ADCC)

Host
- Mechanical barrier limiting contact of tumor with antibodies
- Insufficient vascularization of or shunts and sinusoids in the tumor
- Absorption and degradation of antibodies in lung, liver, spleen by Fc-mediated mechanism
- Qualitative or quantitative insufficiency of the complement system and/or ADCC cells
- Induction of immunosuppressive mechanisms by the antibodies

Thus it was possible to select antibodies which were highly specific for tumor-associated antigens. By means of these antibodies, epitopes with a high degree of tumor specificity for those tumor-associated antigens have been defined immunochemically. Isotope classes of these monoclonal antibodies have been selected which induce strong effector mechanisms and can therefore be regarded as tools for tumor immunotherapy. Radioimmunoscintigraphy studies in immunodeficient mice bearing human tumor transplants, as well as in tumor patients, proved the tumor-specific localization of these antibodies in vivo. Consequently, the concept of targeting drugs onto malignant cells – which had been proposed by Paul Ehrlich [176] and substantiated by Pressman and Keighley [539] with the suggestion of using antibodies as a carrier (a proposal which had first been tested experimentally with ^{131}I-conjugated antibodies by Pressman and Korngold [540], with conjugated aminopterin in L1210 leukemia by Mathé et al. [445], and clinically with ^{131}I-conjugated antibodies [43], and with conjugated chlorambucil [225] in malignant melanoma) – could now be elaborated in detail (see table III). Chemical methods of coupling plant toxins, isotopes and cytostatics to monoclonal antibodies have been developed and these immunotoxins are now being tested preclinically as well as clinically. The continuing clinical studies in immunoscintigraphy, radio-, chemo- and immunotherapy now show us the chances and the limitations of such treatment. The

Table III. History of targeting drugs onto malignant cells

Year	Author	
Proposals		
1900	Ehrlich	'magic bullet'
1948	Pressman and Keighley	antibodies as carriers for cytostatics and radionuclides
Experiments		
1951	Beierwaltes	^{131}I-labelled antibodies for treatment of human melanoblastoma
1953	Pressman and Korngold	^{131}I-labelled antibodies for detection of rat tumors
1958	Mathé et al.	aminopterin-conjugated antibodies for treatment of L1210 leukemia
1972	Ghose et al.	chlorambucil-conjugated antibodies for treatment of human melanoma

Table IV. Considered possibilities for using monoclonal antibodies in tumor therapy

In vivo	Ex vivo
Immunoscintigraphy	immunocytology
	histology
Immunotherapy	
single treatment	bone marrow purging
combination therapy	
effector cells	
mediators	
Radioimmunotherapy	
Chemoimmunotherapy	
Local or systemic treatment	absorption of antigens out of the blood

low amount of antibody localizing in the tumor, and the xenogeneity of the murine monoclonal antibody seem to be the main problems. Murine monoclonal antibodies induce antibodies in the recipient, a part of them being directed to the variable region of the antibodies. This experience is now forcing us either to produce human monoclonal antibodies instead of murine ones or to humanize those murine antibodies by recombinant techniques.

It is the aim of this book to review and discuss critically the state of the art of the different uses of monoclonal antibodies in tumor therapy (see table IV) and to outline the goals to which future research should aspire.

Tumor Antigens

The presence of accessible and operationally specific tumor antigens is an essential and minimal prerequisite for a successful antigen-specific tumor immunotherapy using antibodies. It would, however, be better to have specific tumor antigens. In spite of the indubitable proof of tumor-specific antigens shown by transplantation experiments in chemically induced [23, 201, 257, 366, 538] or virally induced (for review see [392, 506]) experimental tumors, the detection of such tumor-specific antigen in spontaneous autochthonous tumors has proved to be extremely difficult. Exceptions are those 'spontaneous' tumors which are obviously virally induced, such as oncorno-induced lymphosarcoma in cats [626, 669] or HTLV-induced leukemias in man [301, 652, 747]. In these tumors, just as in the virus-induced experimental tumors, viral-associated tumor-specific antigens were defined. Based on experience with experimental tumors, it may be assumed that other human tumor diseases which strictly correlate with viral infection (see table V) are also characterized by the appearance of viral-associated tumor-specific antigens.

Table V. Human virus with oncogenic properties (according to Wyke and Weiss, 1984)

Virus	Tumor
Hepadna, hepatitis B	hepatocellular carcinoma
Herpes, Epstein Barr	Burkitt's lymphoma immunoblastic lymphoma nasopharyngeal carcinoma
Retro, T cell leukemia	adult T cell leukemia-lymphoma
Papova, papilloma	warts cervical cancer laryngeal cancer skin cancer

Table VI. Selected tumor-associated antigens (for details see Stefanini et al., 1985, Melchert and Kreienberg, 1980, Lamerz and Fateh-Moghadam, 1975, Havemann et al., 1986, Bostwich et al., 1984)

	Increased in tumors of
A) Oncofetal antigens	
Carcinoembryonic antigen (CEA)	gastrointestinum, lung, mamma, ovary, testis, kidney, thyroid gland, etc.
Alpha-fetoprotein (AFP)	liver, testis
β-oncofetal antigen	colon, melanoma, endometrium
Fetal sulphoglycoprotein antigen	stomach
Isoferritin	liver, colon, lung, mamma
Carcinoplacental alkaline phosphatase	liver, pancreas
β-S-fetoprotein	colon, mamma
B) Pregnancy associated antigens	
β_1-glycoprotein (SP$_1$)	mamma, lung, gastrointestinum, urogenital organs, chorion
α_2-glycoprotein (SP$_3$)	mamma, lung, gastrointestinum, urogenital organs, chorion
Regan-isoenzyme of alkaline phosphatase	mamma, lung, gastrointestinum, urogenital organs, chorion
C) Hormones	
Human chorionic gonadotropin (HCG)	chorion, ovary, mamma, testis, pancreas, liver, lung, gastrointestinal organs
Human placental lactogen (HPL)	chorion, ovary, mamma, testis, pancreas liver, lung, gastrointestinal organs
ACTH, FSH, STH, neurophysin, TSH	pituitary gland, lung (small cells), stomach (STH), kidney (neurophysin)
Insulin	pancreatic islets, connective tissue (sarcoma)
Gastrin	pancreas, ovary
Calcitonin	thyroid gland, uterus, bladder, breast, lung, prostate
Erythropoietin	kidney, uterus
Parathyroid hormone	parathyroid, breast, kidney, lung, liver parotis, testis
Prostaglandin	thyroid gland, breast
Gastrin releasing peptide	lung

Table VI. (continued)

	Increased in tumors of
D) Enzymes	
Amylase	lung
Acid phosphatase	prostate, CML, breast
Alkaline phosphatase	bone
Histaminase	ovary, thyroid gland
Muramidase	mouse leukemia, gastrointestinal organs, CNS
Hexokinase	breast, lung
Leucin-aminopeptidase	uterus, leukemia
β-glucuronidase	kidney, prostate, uterus, ovary
Prolylhydrolase	liver, breast
Cystine-Aminopeptidase	endometrium, uterus
Sialyltransferase	mamma, uterus, ovary,
Fucosyltransferase	lung, colon, stomach,
Galactosyltransferase	thyroid gland
Enolase	pancreas islet, lung (SCLC), thyroid gland, apudomas
Plasminogen activator	endometrium, ovary
L-dopa-decarboxylase	lung
Terminal deoxynucleotidyl transferase	leukemia (ALL), granulocytes (CGL)
E) Milk proteins	
κ-casein	mamma, bronchus
Lactoferrin	
α-lactalbumin	
F) Metabolites	
Nucleosides	mamma, lung, ovary,
Polyamines	testis, uterus, leukemia
α-2-H ferritin	
β-2-microglobulin	

Instead of tumor-specific antigens in human tumor diseases a huge number of tumor-associated antigens (TAA) were found (see table VI). These belong to oncofetal proteins, enzymes, hormones, pregnancy-associated antigens and differentiation or tissue-specific antigens. TAA are usually produced in limited amounts by the corresponding normal tissue. However, a variety of conditions, e. g. hyperplasia, may substantially increase the production of these or of cross-reacting antigens. The quantitative differences in the amount of these TAA found in tumor cells and normal cells have been widely used for serological tumor diagnosis. The use of polyclonal antibodies against these TAA in tumor immunotherapy was, however, affected by the presence of oncofetal or tumor-associated antigens in serum and in normal tissues.

The hybridoma technique for the production of monoclonal antibodies [370] provided the instrumental device for finding new TAA or new epitopes on tumor antigens and of checking whether those epitopes were also present on normal cells.

Three categories of TAA have been suggested [542] (see table VII), depending on their occurrence or on the occurrence of selected epitopes on TAA in different tumor patients (1), in tumors of similar histological type (2) and in normal tissues (3). Class 1 TAA are tumor specific but restricted to individual tumors. Class 2 TAA are specifically associated with certain

Table VII. Categories of tumor-associated antigens (according to Price et al., 1980)

Category	Definition	Example
Class 1	restricted to autologous tumor cells (individual tumor specific)	blood group A-like antigen in tumor of blood group 0 or B patients; blood group P_1 in tumor of pp genotype patients [83, 259, 277, 304, 348, 746] idiotype of immunoglobulins in B cell lymphoma [461–463]
Class 2	common to histologically similar tumors from several patients, absent in other tumors and normal tissues	O-acetylated GD_3 disialoganglioside (melanoma); virus-associated tumor antigens [126] fucosyl-GM1 with two fatty acids (SCLC) [499]
Class 3	common to tumor cells and various components of normal tissue	nearly all TAA known up to now

types of tumors, and cannot be found on normal tissues. Thus TAA of class 1 and 2 can be called tumor-specific antigens (TSA). The vast majority of TAA, however, belong to class 3, i.e. they can also be found in normal tissue.

Monoclonal antibodies directed against such class 3 TAA can gain tumor selectivity by various means (see table VIII): either through the existence of a significant difference in the amount of TAA produced by tumor cells compared to normal cells or through TAA accumulation in or around the tumor. Another factor is the degree of accessibility of TAA to monoclonal antibodies. In normal cells this accessibility might be hindered either by the cell membrane, the blood-brain barrier or other barriers such as the Lamina propria of glands, whereas in tumors these barriers might be disrupted or without biological significance due to the infiltrative growth of the tumor cells or due to the increased exposure of TAA in the membrane. On the other hand, new barriers can arise in tumors caused by lack of vascularity due to growth of connective tissues, sinusoids and vascular shunts or due to systemic distribution of shed TAA, and these barriers might hamper monoclonal antibodies reaching the tumor.

An increase of selectivity for tumors could also be achieved by raising monoclonal antibodies to epitopes, located in TAA, but more or less expressed only by tumor cells.

Table VIII. Gain in tumor selectivity of class 3 TAA

Mechanism	Example
Quantitative differences in production of TAA between tumor cells and normal cells Local accumulation of shed TAA in and around tumors	oncofetal antigens, i.e. CEA, AFP; ectopical production of hormones and prohormones; tissue-specific enzymes (enolase); membrane-associated antigens
Lack of accessibility to normal cells due to physiological barriers (blood-brain barrier, cell membrane)	expression of GD_2, GD_2 in melanoma cells and in CNS cells; exposure of Thomsen-Friedenreich (TF) antigens in invasive growing mammary carcinoma cells compared to normal cells and milk droplets; epitopes on mucins excreted by pancreas cells and cells of pancreatic ductuli
Selection of epitopes on TAA, the exposure of which are dependent on structural conformation and/or selective for tumors	epitopes on CEA, exposed only in case CEA is bound to solid phase (cell membrane, carrier) (BW 431/26)

Table IX. Selection of tumor-associated antigens detected by monoclonal antibodies

Tumor (immunogen/specificity)	MAb	Antigen (MW)	Reference
Breast	B 72.3	mucin (>100 KD)	Colcher et al. 1981 Nuti et al. 1982
	DF 3	(29 KD)	Kufe et al. 1984
	Ma 5, -7, -9	glycoprotein (280–300 KD)	Major et al. 1987
	F 36/22 HMFG1 HMFG2	mucin-like glycoprotein high molecular weight glycoprotein of human milk	Papsidero et al. 1984 Taylor-Papadimitriou et al. 1981 Arklie et al. 1981
	B 72.3 B 6.2 UCD/ABGI	glycoprotein (220–400 KD) (90 KD) (54–56 KD)	Stramignoni et al. 1983 Schlom et al. 1984 Brabon et al. 1984
Prostate	D 83.21	glycoprotein (8.8 KD)	Starling and Wright 1985
Colon carcinoma	19–9	glycoprotein (200 KD) (sialylated Lewis A)	Koprowski et al. 1981 Herlyn et al. 1982 Magnani et al. 1982
	17-1A	unknown	Herlyn et al. 1982 Atkinson et al. 1982
	NP1-4 35; 73 Col 1-15 BW 431/31 BW 431/26 BW 250/183 F4/2E10	CEA (180 KD) CEA (180 KD) CEA (180 KD) selective CEA selective CEA NCA 95 glycoprotein (35, 42, 90 KD) colon-ovarian tumor antigen (COTA) gastrointestinal cancer associated antigen p21–ras	Primus et al. 1983 Buchegger et al. 1984 Muraro et al. 1985 Bosslet et al. 1985 Bosslet et al. 1987 Bosslet et al. 1985 Drewinko et al. 1986 Pant et al. 1983 Atkinson et al. 1982 Gollick et al. 1985
Ovarian carcinoma	OC 125 $2C_8$; $2F_7$ B 72.3 Tag	glycoprotein (75 KD) glycoprotein (60 KD) glycoprotein (72 KD)	Bast et al. 1981, 1982 Bhattacharga et al. 1985 Thor et al. 1986

Table IX. (continued)

Tumor (immunogen/specificity)	MAb	Antigen (MW)	Reference
Pancreas	DuPan-2	high molecular weight protein	Metzgar et al. 1982
	BW 494/32	carbohydrate structure on mucin (>200 KD)	Bossslet et al. 1987
	AR2-20	(190 KD)	Chin and Miller 1985
	ARI-28	(10 KD)	Chin and Miller 1985
	C-P83	(100 KD)	Schmiegel et al. 1985
	YPan1/-2	mucin-like glycoprotein	Yuan et al. 1985
	C54-0	(glyco)protein (122 KD)	Schmiegel et al. 1985
	C1-N3	(105–135 KD)	
	C1-P83	(110 KD)	
	15.75	glycoprotein (74 KD; 49 KD)	Johnson et al. 1981
	345.134S	glycoprotein (115 KD)	Imai et al. 1982
Melanoma	p97	transferrin (97 KD)	Reisfeld 1985 Brown et al. 1982
	R-24	GD_3	Dippold et al. 1984
	D1.1	O-acetylated GD_3	Cheresh et al. 1984
	OFA 1-2	GD_2	Saito and Cheung 1985
	9.2.27	glycoprotein-chondroitin sulfate proteoglycan complex (MW 250 KD)	Bumol and Reisfeld 1982
	ME 492	melanoma-associated glycoprotein (30–60 KD)	Atkinson et al. 1985
	F11	sialoglycoprotein (100 KD)	Bumol et al. 1982
Lung	MOC-1	antigen on SCLC	de Leij et al. 1985
	TFS-2	antigen on SCLC	Okabe et al. 1984, 1985
	TFS-4	antigen on SCLC	Watanabe et al. 1987
	SM-1	antigen on SCLC	Bernal et al. 1984
	GOOD1	antigen on SCLC	Zimmer et al. 1985
	211	bombesin	Sell and Reisfeld 1985
	11G11	antigen on SCLC	
	B10	100 KD protein (SCLC)	Reeve et al. 1985
	LuCa-2	squamous cell carcinoma	Kyoizumi et al. 1985
	703 D_4	proteins on non-SCLC	Mulshine et al. 1983
	704 A_1	(31 KD)	
Brain	C12; D12	(glyco)protein glioma (180 KD; 88 KD)	Wikstrand et al. 1986
	126	GD_2 (neuroblastoma)	Schulz et al. 1984
Bone	79IT/36		Embleton et al. 1981 Farrands et al. 1983

A variety of MAbs were raised which bound to epitopes more or less selective for certain tumors (see table IX). These tumor-associated epitopes (TAEs) can be composed of protein only, protein combined with carbohydrates, or carbohydrates only. Many different TAEs can exist on one single tumor-associated antigen (TAA). Antibodies reactive to different epitopes located on one TAA, i.e. carcinoembryonic antigen (CEA), can have a totally different specificity on human tissues [69, 90].

This phenomenon is explained by the presence of cross-reacting antigens in certain tissues. These cross-reacting antigens have some epitopes in common with the TAA. But there are epitopes which are unique to a certain TAA and thus not expressed on the cross-reacting antigens. MAbs directed against these epitopes (so-called class V epitopes [389–391]) are very tumor selective and can be used for immunoscintigraphy and/or immunotherapy of cancer.

The best example of the above is the glycoprotein carcinoembryonic antigen (CEA) [243] and its differentiation to cross-reacting antigens [270, 271, 282, 284, 285, 389–391]. These show a considerably amino acid sequence homology to CEA (see table X) by means of monoclonal antibodies [69, 74, 282, 284, 285]. Today we know of several monoclonal antibodies (II–17, II–7, II–10, II–16 [282–285]; MAb 35 [89]; BW 431/31 and BW 431/26 [69, 74]) which specifically recognize CEA but which do not bind to any known cross-reacting antigens.

In view of the complex carbohydrate structure of CEA and in view of the considerable microheterogeneity in its approximately 40 oligosaccharide chains [118, 174, 270, 271], the chance exists that there might be tumor-specific epitopes on CEA. Up to now, however, no differences between CEA produced by colorectal or ovarian, pancreatic, lung and breast cancer could be detected [271].

However, by means of selection Bosslet et al. [74] found the monoclonal antibody BW 431/26 which binds to a CEA-specific epitope which is mainly expressed on CEA bound to the cell membrane or attached to a solid phase. Thus CEA in solution does not react with BW 431/26. Consequently, BW 431/26 is not neutralized by CEA in serum and thus seems to be superior to other anti-CEA antibodies with respect to tumor localization.

CEA is the most prominent TAA detectable in gastrointestinal (GIT) carcinomas. MAbs reactive to CEA-specific protein epitopes already play a clinical role in the field of immunoscintigraphy of colorectal and pancreatic cancer. Besides CEA, other TAAs such as COTA (colon-ovarian tumor antigen) [520], GICA (gastrointestinal cancer associated antigen) [20], and the

Table X. CEA and cross-reacting antigens

Name	Synonyms	Molecular weight	Carbo-hydrate (%)	Similarity to CEA (amino acid composition)	Occurrence
CEA	—	180,000	50–60		colon carcinoma, normal colon
NCA-95*	NCA, NGP	95,000	20–30	+ +	colon carcinoma, neutrophile granulocytes, serum
NCA-55*	NCA, NGP	55,000	?	+ +	colon carcinoma, neutrophile granulocytes, normal epithelium of lung and pancreas, serum
NCA-160	MA	160,000	45–50	+ + + (degradation product of CEA?)	colon carcinoma, normal colon, feces, meconium
CEA_{low}	—	125,000	30–40		colon carcinoma
BGP I	—	83,000	35–40	(+)	bile, serum
NFA I	—	20–30,000	10–15	+ +	feces

CEA	=	carcinoembryonic antigen
NCA	=	non-specific cross-reacting antigen (Kleist et al. 1972)
NGP	=	normal glycoprotein (Mach und Pusztaszeri 1972)
*	=	differentation by Buchegger et al. 1984
MA	=	meconium antigen (Burtin et al. 1973; Primus et al. 1983)
CEA_{low} =		(Hedin et al. 1978)
BGPI	=	biliary glycoprotein I (Svenberg et al. 1979)
NFA-1	=	normal fecal antigen (Kuroki et al. 1981)

oncogene product $p21^{ras}$ [247] have been described and their clinical utility is being investigated. Carbohydrate structures of high molecular weight mucins as well as of glycolipids have been described as being associated with pancreatic carcinomas [71, 432, 433]. MAbs against these structures are at present under clinical investigation for the treatment of pancreatic carcinomas [657].

Small cell lung carcinoma, the most tumorigenic of the 4 lung carcinoma groups, is thought to be of neuroectodermal origin. This assumption was substantiated during the first International Workshop on Small Cell Lung Cancer Antigens (for details, contact Dr. P. C. L. Beverley, School of Medicine, University College of London, University Street, London WC1 E6JJ). Using immunohistological, immunocytochemical and cytofluorometric analysis, antigens on SCLC were classified into 6 main clusters according to their specificity on human tissues. Twelve different MAbs (Cluster 1) were detected showing binding to SCLC and neuronal cells, thus supporting the argument in favor of the neuroectodermal origin of SCLC. Additionally, a variety of antigens were detected expressing epitopes on SCLC and on normal epithelia (Cluster 2), proving that SCLC carries markers which are also present on epithelial cells. The take-home lesson from this workshop was that SCLC expresses a variety of different antigens carrying a myriad of epitopes which have to be classified in more detail in future workshops. Because of the tremendous complexity of the cell surface, an exchange of monoclonal reagents between the laboratories working on this area is vital, in order to bring a little light into the bulk of literature dealing with those MAbs reactive to SCLC lines and tissues [47, 150, 418, 440, 480, 481, 498, 499].

Human non-SCLC contain 3 major histological types: squamous cell carcinoma, adenocarcinoma and large cell lung carcinoma. A variety of MAbs were generated against these tumors showing cross-reactivities with squamous cell or adenocarcinomas derived from tissues other than lung epithelium [110, 288, 712]. This field of research as well as that of SCLC needs a workshop in which different MAbs are exchanged and typed to determine clusters of reactivity, as was done for T lymphocytes.

Human mammary carcinomas can be histologically classified in ductular and lobular carcinomas. Both tumor types express a variety of heterogeneously expressed differentiation antigens whose epitopes are also expressed on the breast ductular system or the lobuli [173]. The most intensively investigated antigenic systems are the structures associated with the milk fat globule membrane which are defined by MAb HMFG2 and HMFG1. The mucins detected by the above-mentioned MAbs have, if they are expressed in normal breast epithelium, carbohydrate structures masking tumor selective protein epitopes which become unmasked in the breast carcinoma tissues. Such an unmasked protein epitope is detected by MAb SM3 which shows a strong binding to breast cancer but not to normal ductular epithelium of the breast [101]. A number of other investigators characterized epitopes located on high molecular weight mucin such as breast-cancer-associated materials which seem to be different from the HMFG system [133, 329, 434].

Melanoma, neuroblastoma, SCLC and brain tumors are derived from neuroectoderm and express glycolipids found in the nerve system.

Ganglioside GD_3 was found to be associated mainly with melanoma [162], GD_2 is expressed in neuroblastoma [655], and Fucosyl-GM1 in SCLC [498, 499]. The role of these ganglioside structures in this type of tumor or in normal tissues is not yet known in detail.

The use of the monoclonal antibody technique in conjunction with achievements in carbohydrate chemistry enabled the detection and structural analysis of carbohydrate structures representing tumor selective epitopes. Most of these tumor-associated epitopes are caused by an aberrant glycosylation of proteins or lipids in tumor cells (see table XI; for review see [265, 266]).

Aberrant glycosylation has been observed in tumors transformed either by DNA and RNA tumor viruses, or chemical carcinogens and in spontaneous tumors, including human cancer. Aberrant glycosylation is presumably caused upon oncogenic transformation by a transforming gene activation leading either to suppressed or repressed glycosyltransferases or an activation or derepression of unusual enzyme(s) with less restricted substrate specificity [197, 263, 711].

Tumor-associated epitopes, resulting either from incomplete synthesis of carbohydrate chains with or without the corresponding precursor accumulation or from neosynthesis of carbohydrate chains, were detected in a number of tumors (see tables XII, XIII).

Research on oncogenes only contributed slightly to the detection of TAA with high tumor selectivity. Thus monoclonal antibodies to ras, sis,

Table XI. Epitopes caused by aberrant glycosylation in tumor cells (according to S. Hakomori, 1985)

In glycoproteins
- Increased carbohydrate branch (GlcNAc-mannosyl core structure of asparagine-linked oligosaccharide is increased)
- Increased carbohydrate density (0-glycoside mucin type oligosaccharide chain)
- Changes in peripheral region (as in glycolipids)

In glycolipids
- Incomplete synthesis with or without precursor accumulation
- Neosynthesis

Table XII. Structural entities with tumor association in man

Block in synthesis of	Precursor accumulation of	Tumor	Author
GM$_3$ (hematoside)	LacCer (Galß1-4Glc-Cer)	brain	Kanazawa and Yamahawa 1974
Gb$_3$, Gb$_4$, Gb$_5$ (globotriasylCer; globoside, Forssmann)	LacCer	kidney	Karlson et al. 1974
GD$_{1a}$, GD$_{1b}$, GT$_{1b}$	LacCer	colon	Siddiqui et al. 1978
	GD$_3$	brain	Eto and Shinoda 1982 Schengrund et al. 1985
	GD$_3$	melanoma	Nudelman et al. 1982 Pukel et al. 1982
	GD$_3$	ANLL	Siddiqui et al. 1984
		T cell	Merritt et al. 1987
	GD$_2$	brain	Cahan et al. 1982 Schengrund et al. 1985
	Gg$_3$ (asialo GM$_2$)	Hodgkin's	Kniep et al. 1983
	Gb$_3$ (globo-triasyl-ceramide)	Burkitt's lymphoma	Nudelmann et al. 1983
Blood group A, B, H (0)	I (MA) (Galß1-4GlcNacß1-6)	lung	Hirohashi et al. 1984
Blood group M, N	T-F antigen	breast, colon, stomach, bladder	Springer et al. 1983, 1985 Ohoka et al. 1985

erb-B src and myc oncoproteins stained sections of tumor tissue as well as of normal tissue [181]. On the other hand, MAb specific for the neu-oncoprotein could inhibit proliferation of neuroblastoma cells specifically expressing that protein [170].

It is obvious that monoclonal antibodies can only bind to those tumor cells which expose the antigen. The number of sites available on cells of experimental tumors [359] or on human tumors [229] has been shown to be mostly in the range of $2.5 \times 10^4 - 7 \times 10^6$ per cell, but may also be significantly lower. Variants of tumor cells with low or no TAA expression escape the binding of specific antibodies and consequently escape any secondary cytotoxic attack. In view of the extreme environmentally influenced, morphological, antigenic and biological heterogeneity of tumor cells within a tumor [173, 193, 205, 212, 310, 311, 743] and between the

Table XIII. Structural entities with tumor association in man (nomenclature according to the IUPAC-IUB Commission on Biochemical Nomenclature 1977 and according to Hakomori (1981) and Hakomori and Kannagi (1983))

Neosynthesis of	Tumor	Author
Fuc Cer (Fucα1-Cer)	colon	Watanabe et al. 1976
X-glycolipid/ polyX-glycolipid (fucosyllactosaminolipid)	colon lung	Brockhaus et al. 1982 Hakomori et al. 1982 Huang et al. 1983
Sialosyl-Le (Fuc (αGal) GM_1)	colon stomach pancreas	Magnani et al. 1981 Magnani et al. 1982 Koprowski et al. 1981 Fukushima et al. 1984
Fucosyl GM_1 with 2 hydroxy fatty acids	lung (SCLC)	Nilsson et al. 1986
Gb_4 (globoside)	myelomonocytic leukemia	Lee et al. 1982
Difucosyl Y_2 III $V^3Fuc_2nLc_6$	colon lung, breast	Hakomori et al. 1984 Fukushi et al. 1984
Sialosylfucosyl Y_2 $III^3 V^3 FucIV^3 NeuAc$	colon lung, breast	Fukushi et al. 1984
Difucoganglioside ($VI^3 NeuAcV^3 III^3 Fuc_2 nLc_6$)	stomach, colon lung, breast, kidney	Fukushi et al. 1985
Globo series (Fucα1-2Galß1-3GalNacß1-3)	breast teratocarcinoma	Bremer et al. 1984 Kannagi et al. 1983
Fucosyl-GM_1	small cell lung carcinoma	Nilsson et al. 1984
Blood group A-like antigen in tumor of blood group 0 or B patients	stomach colon liver	Breimer 1980 Häkkinen 1970 Hattori et al. 1981 Yokota et al. 1981 Hirohashi et al. 1984
Forssman antigen	stomach colon lung	Hakomori et al. 1977 Yoda et al. 1980 Taniguchi et al. 1980 Yokota et al. 1981
Blood group P_1 in tumor of pp genotype patients	stomach	Levine et al. 1951 Kannagi et al. 1982
0-acetylation of sialic acid in gangliosides	melanoma	Cheresh et al. 1984
Asialo-3fuc-NAc lactosamin	AML, ALL	Stockinger et al. 1984

primary tumor and its metastases [188, 209], the selection of TAA-negative tumor cells through the application of monoclonal antibodies is preprogrammed. This is valid for the application of antibody alone as well as of toxins or cytostatics, conjugated to the antibody. It might not, however, be the case for isotopes linked to the antibody, because the radiocytotoxicity of these isotopes might not be restricted to the tumor cell the respective antibody is bound to, but can also affect neighboring cells including TAA-negative tumor cells.

On the other hand, it has been shown that much of the antigenic heterogeneity of tumor cells might be due to a dynamic process of antigenic modulation, i.e. the turning on and off of the expression of given tumor antigens by factors such as cell cycle, environmental milieu, or spatial configuration of lesions [645, 646].

The problem of antigenic modulation and heterogeneity in the expression of TAA could be approached through the use of biological response modifiers or drugs that intensify and increase the expression of TAA on the surface of carcinoma cells. Thus, treatment of human breast or colon carcinoma cells with α- or γ-IFN dose-dependently and transiently increases the tumor-specific surface expression of TAA as evaluated by monoclonal antibodies [33, 240, 252–254]. These studies clearly showed that the α- or γ-IFN mediates an increase in the percentage of cells expressing the TAA as well as an accumulation of more TAA per cell. Consequently, concomitant administration of IFN and monoclonal antibody may increase the efficacy of monoclonal antibodies in the detection and treatment of carcinoma lesions [254].

Apart from the concentration, the accessibility of the TAA is of decisive importance for the antibody. Steric hindrance of glycolipid TAA by other membrane-located glycolipids and glycoproteins [265], membrane fluidity and the relatively slow diffusion constants of membrane-bound antigens may prevent the proper orientation of the epitope and thus its binding with the antibody [238].

To summarize, according to our updated knowledge, the occurrence of tumor-specific antigens seems to be a rare event which is restricted to tumors of certain individual patients or to virally induced tumors or to melanomas. Most of the numerous tumor-associated antigens or epitopes on those antigens which have been detected and analyzed by biochemical and immunological methods, including monoclonal antibodies, are not tumor-specific. However, they gain at least an operational tumor specificity by quantitative differences in the expression by tumor cells, compared to normal cells. Moreover, due to tumor growth, anatomical conditions might arise which allow monoclonal antibodies to reach TAA on tumor cells, but not on normal cells. The degree to which any defined TAA can be used as a target for monoclonal antibodies depends on additional parameters such as degree of exposure and homogenicity and stability in the expression of TAA in the tumor nodule (see table XIV).

In most cases we know little or nothing about the TAEs on the TAAs or their function. At the moment our research is confined to a phenomenological approach, looking for specific structures definable by MAbs.

Future research will hopefully produce further insight into the slight structural differences between tumor cells and normal cells, the reasons for these differences and their functional relevance. This insight will then make the search for tumor-selective monoclonal antibodies a more rational strategy.

Table XIV. Features of TAA which are decisive for diagnostic or therapeutic tumor localization of monoclonal antibodies

— Difference between the amount of TAA expressed by tumor cells and normal cells
— Absolute number of TAA molecules per tumor cell surface
— Relative number of tumor cells expressing TAA compared to TAA-negative tumor cells
— Homogenicity in the distribution of TAA-positive tumor cells within a tumor
— Variation over time in the expression of TAA by tumor cells
— Degree of exposure of epitopes of membrane-associated TAA for monoclonal antibodies (movement; lability in the membrane; steric hindrance)
— Secreted TAA versus TAA representing membrane components
— Reactivity of soluble versus cell-membrane-bound TAA to monoclonal antibodies
— Accumulation of TAA in/around the tumor
— Fate of antibody-TAA complexes on the cell membrane (internalization; capping; shedding)
— Size of the tumor
— Degree of accessibility in vivo of TAA on tumor cells compared to normal cells for monoclonal antibodies
— Degree of vascularization of the tumor
— Degree of infiltrations, fibrotic reactions, necrosis in the tumor mass

Generation, Selection, Properties and Quality Control of Murine Monoclonal Antibodies

Generation and Selection

As already mentioned in the previous chapter, the detection of murine monoclonal antibodies of sufficient operational specificity for tumors is more or less like gold digging: we know what kind of matter we are looking for, but we do not know whether and, if at all, to which degree we can find it.

Basically, the Köhler and Milstein technique [370] is generally employed, whereby in the meantime a considerable number of myeloma cells have been developed for hybridization (see table XV). These myeloma cells must fulfill several criteria: first, they must be sensitive to media containing hypoxanthine, aminopterin and thymidine (HAT) [153, 636] allowing a selection of hybrids after fusion. This is achieved by growing the myeloma cell line in azaguanine, which selects cells that are hypoxanthine-guanine-phosphoribosyl transferase deficient. Thus the cells lack the enzyme necessary to rescue them from aminopterin toxicity. A second requirement is the inability to produce their own rearranged immunoglobulin. In our work, we mainly use the myeloma cell lines X63Ag8/653 and SP2/0-Ag-14 which guarantee sufficient fusion and cloning efficiency. Supernatants of hybrids of these myeloma cells with spleen cells of mice repeatedly immunized with selected tumor material (see fig. 1) are tested for specific binding to tumor cells in bioptic, surgical or xenografted human tumor tissue sections. Hybrids secreting antibodies with specific binding to tumors and no obvious binding to blood cells and normal tissues are selected for further development and antibody characterization (see fig. 1). This primary screening (which is similar to that of de Leij et al. [150]) on human tumor tissue prevents us from selecting an irrelevant specificity.

Former use of established tumor cell lines as antigen targets for primary screening misled us to select a battery of false-positive antibodies. Perhaps we even missed identifying interesting specificities because of false-negative results. These mistakes were caused by differences in antigen

Table XV. Myeloma cell lines commonly used for hybridization (for review see Edmond et al., 1986; James and Bell, 1987)

Name	Origin	Strain	Producer of
Mouse			
MOPC-21	plasmacytoma	Balb/c	IgG_1; IgG_{2b}
MPC-11			
Variants of MOPC-21			
P_3-NS1-Ag4-1	plasmacytoma	Balb/c	L-chain
P_3-X63-Ag8/653	plasmacytoma	Balb/c	—
SP2/0-Ag-14	plasmacytoma	Balb/c	—
Rat			
Y3-Ag-123	myeloma	Lou/C	K-chain
YB2/0	myeloma	Lou/c/Ao	—
Man			
U-266	myeloma		IgE (λ)
HFB1	myeloma		—
RH-L4	lymphoma		—
GM 1500	LCL		IgG_2 (K)
ARH 77	LCL		IgG_1 (K)
W1-L2	LCL		IgM (K)
W_1-L_2-727	LCL		IgG (K)
MC/Car	LCL		—
CRL-1484	LCL		—
HS-Sultan	LCL		—
GK-5	LCL		K-chain

expression in tumor cell lines compared to xenografted or even surgical material. Thus (in extension of the work of Egan and Henson [175] and Tsuchida et al. [702]) we were able to show by a battery of monoclonal antibodies that cell-membrane-associated tumor antigens in surgical material disappear after cultivating these tumor cells and reappear after xenografting the cultured cells [65, 66].

Up to now, we have developed [64, 66, 67, 69, 71, 74] a set of monoclonal antibodies which are directed against epitopes of CEA, pancreatic carcinoma and bronchial carcinoma, and which dispose of at least operational specificity with respect to their in vivo usage for tumor diagnosis and therapy (see table XVI).

This set of antibodies was selected because two essential conditions were fulfilled: first, the antibodies do not react, or react only marginally,

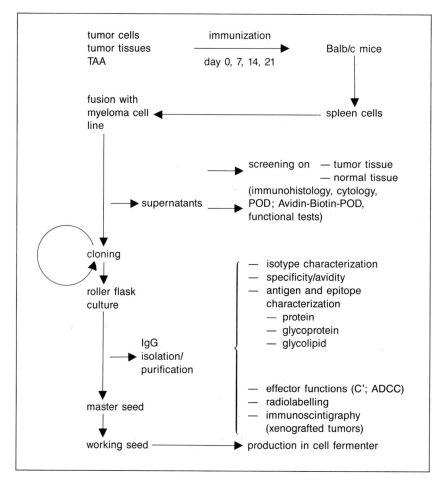

Fig. 1. Selection of monoclonal antibodies.

to normal tissue blood cells or plasma proteins; second, the epitopes of the respective antigens are expressed on the cell membrane of tumor cells. Thus after radiolabelling, when injected i.v., the antibody is able to localize specifically a subcutaneously grafted tumor expressing the corresponding antigen in nude mice. The degree of localization is either calculated by evaluation of the localization index [473] and/or by immunoscintigraphy (see fig. 2). Only those antibodies which concentrate sufficiently on the tumor site can be considered for further development in tumor therapy.

Table XVI. Monoclonal antibodies (Behringwerke) against tumors in clinical trials

Clone No.	Isotype	Specificity (antigen)	Localization index (xenograft)	Immunological function	Reference
BMA 431/31	IgG$_1$	CEA (epitope not on NCA; MCA)	5.0–20.0	—	Bosslet et al. 1985
BMA 431/26	IgG$_1$	CEA (exposed only on solid phase/membrane-bound CEA; epitope not on NCA; MCA)	3.0–5.0	—	1987
BMA 494/32	IgG$_1$	pancreas carcinoma (highly differentiated) > 200 kDa membrane-associated glycoprotein	2.5–3.0	ADCC (medium) inhibition of cell function	1986
BMA 494/32	IgG$_{2a}$ (switch mutant)	pancreas carcinoma (highly differentiated) > 200 kDa membrane-associated glycoprotein	2.5–3.0	ADCC (high) inhibition of cell function	1987
BMA 495/36	IgG$_3$	universally distributed carcinoma-associated antigen 200 kDa glycoprotein	5.0–30.0	ADCC C'-activation	1986
BMA 495/36	IgG$_1$ (switch mutant)	universally distributed carcinoma-associated antigen 200 kDa glycoprotein	5.0–30.0	—	1987
BMA 557	IgG$_1$	non-small cell lung carcinoma	2.5–4.0	—	
BMA 575	IgG$_1$	small cell lung carcinoma	2.8–3.5	—	
BMA 621	IgG$_1$	lung carcinoma (small cell, large cell, adeno, squamous)	3.2–4.5	—	

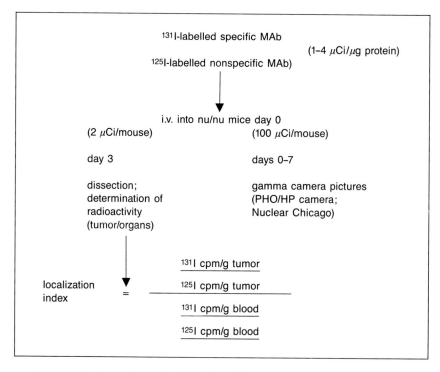

Fig. 2. Experiments in xenografted nu/nu mice.

The degree of localization depends on the amount and the degree of availability of the antigen exposed by the tumor cell, the number of tumor cells exposing the antigen and the quality, especially the specificity and the avidity, of the antibody. Tumor localization of monoclonal antibodies can be supported by increased tumor vascularization, increased permeability of newly formed vessels and the lack of lymphatics in tumors [230–234] and can be hampered by shedding or secretion of the TAA by the tumor cell [670].

As has already been discussed (see p. 5 ff.), the number of sites available on cells of experimental tumors [359] or on human tumors [229] has been shown to be mostly in the range of $2.5 \times 10^4 - 7 \times 10^6$ per cell, but may also be significantly lower. Since the number of antibodies which will bind to the tumor is directly correlated to the concentration of the TAA at the tumor site [359], monoclonal antibodies with specificity for TAA of low concentrations are less suitable for localization. Localization is additionally impaired by TAA which are secreted or shed by living cells or distributed from dead cells into the extracellular fluid. Extracellular TAA may neutralize the antibody either in the immediate vicinity of tumor cells or at a distance from them [461, 487, 488, 490, 670]. By contrast, those antibodies which react to epitopes on TAA being preferentially exposed when the TAA is localized in the cell membrane have advantages for localization. This is for instance the case with our antibody BW 431/26, which binds to a CEA epitope which is mainly exposed when CEA is located in the cell membrane or absorbed to a solid phase [74].

After the monoclonal antibody has bound to the cell-membrane-associated TAA, the resulting immune complex might be shed [258, 569–572, 729] and therefore be inefficient in targeting. On the other hand, internalization of the immune complex by the tumor cell may occur, which enables or improves the cytotoxic effect at least for conjugates of cytotoxic agents and antibodies. The localization of antibodies may additionally be impaired by cell membrane constituents adjacent to the epitope-bearing TAA.

In the case of glycolipid TAA, for instance, other glycolipids in the vicinity may cause sterical hindrance for the antibody to bind the epitope [265]. Another factor could be membrane fluidity, which may prevent the proper orientation of the epitope for binding with the antibody (see p. 5 ff.).

Affinity

The phrase 'antibody affinity' is used to describe the binding constant of the antigen-antibody reaction. If the reaction is simple, the term 'intrinsic affinity' is used. If the reaction is complex, the term 'functional affinity' has been applied [256, 517]. The interaction between antibodies and the cell membrane TAA has been regarded as a typical fluid phase mono- less a bivalent reaction [358, 359]. Thus it is not surprising that IgG antibodies and their F(ab) and F(ab')$_2$ fragments have nearly identical affinity constants [358] in binding to the corresponding cell-membrane-associated antigen. Moreover, at a given antigen concentration, the affinity constant of an antibody determines the amount that can bind to target cells. The lower the affinity of the antibody, the less antibody binds to the tumor [537]. And in the case of low affinity even the fraction that binds is rapidly lost from the tumor [27]. In our experience, as also in the experience of

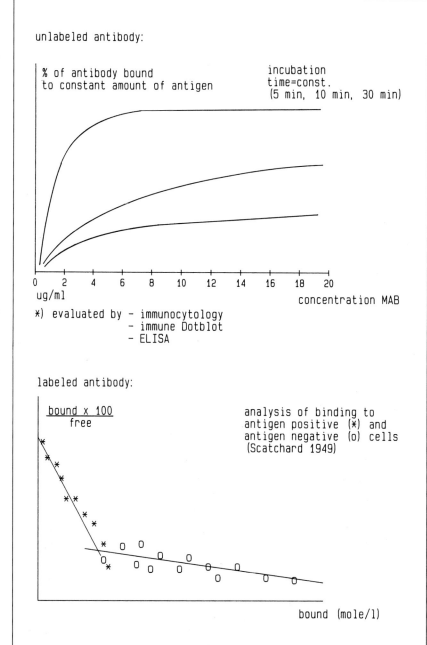

others [359], the use of antibodies having binding constants of less than 10^8 l/mol is probably less suitable for tumor localization, whereby about 10^6 antigenic binding sites per cell are presupposed. As avidity of the monoclonal antibody is one of the essential parameters for binding to the tumor site, its improvement and evaluation is considered within our screening system. Immunization of splenic donor mice is repeatedly carried out with fresh or fixed material of one type of tumor. With this procedure we intend to boost the immune reaction against epitopes on TAA common to and selective for this specific type of tumor (TAA of class 2 or 3).

Tumor material used for immunization may consist of established human tumor cell lines or freshly isolated tumor cells or tumor tissue homogenates or tumor cell membrane preparations from patients. Tumor material may be fixed with formaldehyde or by other means immediately after surgical excision. The use of fresh tumor material for immunization increases the chance of getting monoclonal antibodies with higher specificity for TAA [594, 670, 671]. If possible, TAA are enriched out of the tumor material, as the use of pure TAA preparations for immunization also improves the specificity of the resulting monoclonal antibody [52, 53, 56]. Thus glycolipids are extracted from fresh tumor material by conventional extraction methods. Immunization with TAA is started in a mixture with an adjuvant followed by at least 5 booster injections with or without adjuvant admixture. Boosting the immune response is known to increase the avidity of the resulting monoclonal antibodies [670; 671]. We measure avidity by using two test systems (see fig. 3). In a semiquantitative way, the percentage of antibody bound to a constant amount of antigen in a given period of time is estimated by immunofluorescence in Terasaki plates [64], immunodot assay [280] or by other adequate techniques. This value is compared to the value of antibodies of similar specificity. Thus a relative estimation of the avidity of the antibody can be performed.

Fig. 3. Estimation of affinity of TuMabs. *a* The affinity of unlabelled TuMabs is estimated in relation to a reference antibody by a double antibody immunofluorescence technique. The amount of cellular antigen and the incubation time are constant; the test is performed with an increasing amount of antibodies. In the case of high affinity a high percentage of small amounts of antibody binds to the cellular antigen within the given time. Here the results of 3 antibodies endowed with different affinities are demonstrated. *b* Affinity of TuMAb can be quantitatively measured by Scatchard plot analysis after radiolabelling.

Subsequently, the affinity of selected antibodies to bind to fresh unfixed or formaldehyde-fixed tumor cells is quantitatively evaluated according to Walker [722], based on original papers by Scatchard [595] and Berson [51]. For this procedure the respective monoclonal antibody has to be labelled, various amounts of labelled antibodies are mixed with various amounts of antigen, and the relationship between bound and free antibody is plotted versus the bound one. The Scatchard plot analysis provides an objective measurement of the affinity constant, being mostly in the range of 10^8 l/mol, and rarely of the order of 10^{10} l/mol, as has been found with our monoclonal antibody BW 431/26 (directed against an epitope on CEA) [74]. It is questionable whether the avidity of the antibody can be further increased by selection of clones producing high-affinity variants. An alternative would be site-directed mutagenesis to induce changes in the amino acid sequence in the framework adjacent to the complementary determining region of the variable region. For instance, using genetic engineering methods in the removal of two charged residues at the periphery of the complementary region, the affinity of the antibody for its antigen was increased more than 8-fold, and the ability of this antibody to cross-react with closely related antigens was significantly reduced [576]. The drastically increasing knowledge in the genetic engineering of murine and human antibodies may indeed open up the possibility of improving the affinity of selected antibodies to a degree sufficient for tumor therapy.

Fc-mediated Functions

While binding to the epitope, the antibody induces effector functions via the constant part of its heavy chain forming the Fc part. The degree to which effector functions are induced depends on the isotype of the monoclonal antibody, i.e. on the structure of the constant part of the heavy chains (see table XVII).

In this respect one must consider that, in a number of cases, murine immunoglobulins seem to activate the human complement and human Fc receptor-bearing cells much less effectively than murine cells or rabbit complement [151, 313, 344, 441, 673].

The effector functions are displayed by activation of the complement and by the binding to the Fc receptor of various cells, including NK cells, macrophages, granulocytes and thrombocytes (reviewed by Burton [106] and Jones et al. [340]). Cell membrane-bound antibodies may thus mediate

Table XVII. Effector functions of mouse IgG subclasses with human complement or Fc receptors (compared on the basis of data in Anderson and Looney, 1986; Burton, 1985; Jones et al., 1985; Masui et al., 1986; Kaminski et al., 1986; Houghton et al., 1985)

		IgG_1	IgG_{2a}	IgG_{2b}	IgG_3
Complement activation		—	+/++	+/++	+++
Fc receptor binding human cells					
FcRI (72 KD)	monocytes	+/++	++	—	++
FcRII (40 KD)	monocytes neutrophils eosinophils platelets B-Ly	+/++		+	
FcRX (50–70 KD)	neutrophils eosinophils macrophages NK-cells K-cells LgLy		++	+	

either complement-induced antibody-mediated cytolysis (CMC) or antibody-dependent cellular cytotoxicity (ADCC), inflammatory reactions and/or phagocytosis by the respective Fc receptor-bearing cell. Studies of ADCC with murine antibodies and human effector cells suggest that IgG_{2a} and IgG_3 are the most effective murine heavy chain isotypes [151, 344, 365, 673].

Human complement fixation is best achieved by murine IgM and murine IgG_3 [313, 409, 721]. All these effector functions may play a decisive role in the immune surveillance of infections, parasitic invasions and, last but not least, in tumor growth control (for review see [604]). On the other hand, the binding of antigen-antibody complexes to Fc receptor-bearing cells of the reticuloendothelial system is the essential mechanism by which immune complexes are cleared from the blood and the tissues. This clearing might be supplemented by complement factors which bridge immune complexes to complement receptors of cells of the RES (for review see [600, 601]).

Immunoglobulin aggregates and even monomers seem to be metabolized by the reticuloendothelial system in a way similar to immune complexes. As a result, effector functions induced by monoclonal antibodies have two sides: one is the essential mechanism to induce complement-mediated cytolysis of tumor cells and antibody-dependent cellular cytotoxicity for tumor cells which are desired in tumor immunotherapy. The other side is the uptake of the injected monoclonal antibodies by the reticuloendothelial system via Fc receptor binding. Since the property of Fc-mediated effector functions is supposed to be an essential prerequisite for the efficacy of monoclonal antibodies in tumor immunotherapy, one must accept the Fc receptor-mediated non-antigen-specific uptake by the RES.

In the case of radioimmunoscintigraphy, radioimmunotherapy or chemoimmunotherapy of tumors, however, the uptake via Fc receptor has to be minimized to enable exact tumor imaging and to prevent any other tissues from being damaged by the immunoconjugate. Minimizing is the selection of a (murine) isotype (IgG_1) with reduced effector function activity or proteolytic cleavage of the Fc part from the antibody molecule. The resulting Fab or $F(ab')_2$ fragments have retained their epitope-binding activity which, in the case of cell-surface-bound epitopes, is of an affinity very similar to that of the intact IgG [359], but they have lost either a part ($F(ab')_2$) or all (Fab) of their effector functions (for review see [600-602]).

F(ab) and $F(ab')_2$ fragments, unlike the entire Ig molecule, have been said to be endowed with an increased transcapillary passage and diffusion in extracellular space [238, 718]. Quantitative data have mostly been elaborated with iodinase-sensitive iodine-labelled monoclonal antibodies, but could not be reproduced with stably ^{111}In-labelled antibodies, at least in the case of $F(ab')_2$ fragments [307, 667].

F(ab) fragments bound to tumor cells are less susceptible to endocytosis or shedding because monovalent fragments do not cap [238]. In those cases where there is no shedding or increased internalization, tumor-cell-bound-antigen-antibody-immune-complexes are catabolized between 6 and 24 h after injection. This catabolization has been found to be independent of the IgG class and reactive immunoglobulin fragment used [640, 641].

Thus F(ab) and $F(ab')_2$ fragments might be of greater use to intact IgG in the function of a target-specific carrier for radionuclides or cytotoxic drugs. In this respect it is surprising that the proportion of the total administered amount of ^{131}I-labelled $F(ab')_2$ molecules that localizes in a tumor is much smaller than that of the parent IgG [235-237, 720]. There are indications, however, that this difference is due to higher susceptibility to deiodinases of ^{131}I-labelled $F(ab')_2$ than ^{131}I IgG. Higher susceptibility of ^{131}I-labelled F(ab')$_2$ to deiodinases may also at least in part simulate an increased clearing rate of F(ab')$_2$ fragments. Significant indications for this have been elaborated by Hnatowich et al. [307], Khaw et al. [361], Epenetos et al. [182] and Steinstraesser et al. [667] with ^{131}I-labelled monoclonal antibodies.

It is obvious that the control of the effector functions by adequate test systems is an essential prerequisite for producing monoclonal antibodies of constant quality to be used either in immunotherapy or in immunoscintigraphy, radioimmunotherapy or specific chemoimmunotherapy (immunotoxins, immunocytostatics) of tumors.

Quality Control

Production of monoclonal antibodies by cell culture techniques in batch or continuous processes up to 500-1,000 mg antibody protein/l culture medium is now possible by using different cell culture systems (for review see [423]). Thus in principle a reproducible production of monoclonal antibodies for experimental and clinical trials can take place.

Where monoclonal antibodies are produced for use on human beings one must be sure that they are safe for the patient, while at the same time containing and maintaining their selected specificity and activity throughout the production process to guarantee a beneficial effect for the patient.

Accordingly, the Food and Drug Administration (FDA) in the USA [196] as well as the European authorities have elaborated points to consider when controlling the quality and the safety of monoclonal antibodies (see table XVIII).

Methods already established and used for the control of purity, stability, biochemical characterization and safety of polyvalent immunoglobulin preparations are also used for the control of monoclonal antibodies after adequate adaptation and optimization (for review see Seiler et al. [605b]; moreover, see table XIX).

Special attention is given to viral contamination, since it is known that all murine myeloma cell lines effectively used for hybridization with specifically sensitized lymphocytes exhibit morphological signs of infection with type A and type C viral particles [31, 32, 590, 697, 732]. Even release from the hybrids of large numbers of C-type particles has been observed [732], whereby some were found to be infectious for human embryonic lung fibroblast cells [732].

Table XVIII. Quality and safety of monoclonal antibodies – points to consider (FDA 1987)

Characterization and definition of
- immunogen
- procedures (immunization; screening; cloning)
- fusion partners (immune cells; myeloma cells)
- seed system (master seed; working seed)

Definition of production procedure
- ascites; cell culture procedure
- purification (salt fractionation; size-, affinity-, ion exchange-chromatography)

Quality control
- genetic stability (hybrid mutations)
- viral contamination
  - murine by MAP test, S^+L^- test, MCF test
  - human (EBV, CMV, retrovirus, hepatitis and others)
- sterility (incl. mycoplasma)
- purity (extraneous Ig, contaminants) and homogeneity
- polynucleotide contamination (<10 pgm DNA/dose)
- stability (final preparation)
- specificity, avidity and function

Safety control
- toxicology
  - two species, if possible, in one of which specific antigen is present
  - highest clinical dose and multiple of it
- endotoxins
  - pyrogenicity, Limulus cell agglutination
- test for cross-reactions
  - in vitro
  - in vivo (if possible)
- dosimetry (preclinically) in case of radiolinked MAb

Moreover, mice serving as donors for specifically sensitized lymphocytes or for ascites production might be infected by viruses which are assumed to be pathogenic in man (table XX) or where pathogenicity for man is under discussion (table XXI).

Consequently, the producer or manufacturer must prove that the monoclonal antibody preparation is free of specific murine and human viruses. In addition, the purification procedure should guarantee that the amount of contaminating DNA, injected per dose of monoclonal antibody, is lower than 10 pg.

To be on the safe side, a special technique has been developed by us to inactivate envelope viruses, including retroviruses possibly contaminating

Table XIX. Monoclonal antibodies against tumors – quality control

Technique	Question	Requirements
Kjehldahl (micromethod)	protein content	1 mg/ml (or other concentrations)
Microzone electrophoresis	purity	~ 100% γ-globulin
Ouchterlony/immune electrophoresis	isotype	IgG_1, IgG_2 or IgG_3 L-chain (K or λ)
SDS-polyacrylamid gel electrophoresis	homogeneity	homogenous γ-globulin prepar. or $F(ab')_2$-fragment
HPLC	homogeneity/ aggregates	~ no aggregates (IgG-dimers)
Thin layer chromatography and HPLC	free ^{131}I or free ^{111}In	no free radionuclide
Autoradiography after SDS-PAGE under non-reducing conditions	degradation of antibody by radiation	no degradation detectable
Immunohistology/cytology (semi-quantitatively on fixed tumor cells)	deterioration of variable region by conjugation	confirmation of specificity; no reduction of specificity and titer

the monoclonal antibody preparation [68]. This technique is based on the knowledge that detergents can inactivate envelope viruses. The special invention was the type of detergent selected and the special conditions for treatment of the antibody while not affecting the antibody function.

Before application in man, possible toxic side effects induced by the final monoclonal antibody preparation have to be tested by adequate toxicological tests in at least two species (see table XVIII). These safety tests in mice and rabbits give only limited information: one of the main questions connected with in vivo usage of monoclonal antibodies directed against TAA – i.e. the degree and the in vivo relevance of the binding of monoclonal antibodies to TAA expressed by normal cells and tissues of humans – cannot be adequately answered by these experiments. Even primates could not sufficiently substitute for man to adequately answer this question. Thus, the very correct and very extensive immunohistological and immunocytological investigations on all kinds of human normal cells and tissues have to be the basis from which we estimate and predict in vivo cross-reactions to normal tissues of monoclonal antibodies in man.

Table XX. Safety requirements for monoclonal antibodies: murine viruses — group I (assumed to be pathogenic to man) (according to the FDA, 1987)

Virus	Species affected
Hantavirus (hemorrhagic fever with renal syndrome)	M, R
Lymphocytic choriomeningitis virus (LCMV)	M
Rat rotavirus	R
Reovirus type 3 (reo 3)	M, R
Sendai virus	M, R

M = mouse
R = rat

Table XXI. Safety requirements for monoclonal antibodies: murine viruses — group II (under discussion as being pathogenic to man) (according to the FDA, 1987)

Virus	Species affected
Ectromelia virus	M
K virus (K)	M
Kilham rat virus (KRV)	R
Lactic dehydrogenase virus (LDH)	M
Minute virus of mice (MVM)	M, R
Mouse adenovirus (MAV)	M
Mouse encephalomyelitis virus (MEV, (Theiler's or GDVII)	M
Mouse hepatitis virus (MHV)	M
Mouse rotavirus (EDIM)	M
Pneumonia virus of mice (PVM)	M, R
Polyoma virus	M
Rat coronavirus (RCV)	R
Retroviruses	M, R
Sialodacryadenitis virus (SDA)	R
Thymic virus	M
Toolan virus (HI)	R

M = mouse
R = rat

In this connection, the crucial question which must be answered is whether the respective TAA is more strongly exposed on tumor cells than on normal cells for the in vivo access of monoclonal antibodies. The final proof of the safety of any monoclonal antibody for immuno-, radio- or chemoimmunotherapy can only be given clinically after adequate immunoscintigraphy studies have been carried out which will enable an evaluation of tumor-specific versus unspecific localization of the respective monoclonal antibody. In the case of ('unspecific') drastically increased localization of any monoclonal antibody in tissues such as bone marrow, liver, kidney or other normal tissues, possible side effects in the various organs induced by these antibodies have to be balanced with the advantage for the patient.

One of the essential requirements for the clinical use of monoclonal antibodies is the guarantee of the reproducible quality of monoclonal antibodies. This means that one has to be sure that the selected hybridoma clone, independent of the culture passage, always produces the same quality of monoclonal antibody with respect to its biochemical characteristics, its specificity and its effector functions.

A change in effector functions of a selected antibody might be caused by minor changes in the secondary or tertiary structure of the Fc part, due to genetic instability of the hybrid. In addition, isotype switch variants may also occur. In our own investigations with the hybrid producing MAb BW 494/32 (an IgG_1 antibody recognizing a carbohydrate epitope on a mucin produced by pancreatic carcinoma of differentiated type [71]), we investigated the frequency of the following 3 types of spontaneous hybridoma variants arising under in vitro culture conditions out of a single cell of the 'wild' parenteral BW 494/32 hybridoma cell line [72]:

– idiotype loss variants, i.e. hybridoma cells, secreting intact MAb of IgG_1-K isotype (identical to the MAb 494/32 but without binding capacity to the specific epitope),

– spontaneous non-producers, i.e. hybridoma cells which have spontaneously stopped secretion of MAb BW 494/32 or any other murine IgG, and

– isotype switch variants, i.e. hybridoma cells secreting a MAb with an identical idiotype (identical variable region) but a changed isotype (different constant region).

The production of specific antibody by single-cell-cloned hybridoma cells was tested by using the immunodot assay [280] with semipurified antigen on the solid phase. Moreover, the isotype was evaluated also using the immunodot assay or a highly sensitive ELISA (sensitivity 0.1-1 ng/ml) with isotype specific antisera and a double enzyme amplification system [336, 663]. Starting statistically from one parenteral cell per well, this cell clone was passaged 4, 15 and 30 times. After limiting dilution the supernatants of the individual cells of the respective culture passage were analyzed (fig. 4). A considerable number of idiotype loss variants (17%) and of spontaneous non-producer variants (16%) were detected after only 4 culture passages, whereas isotype switch variants arose in a frequency of 2×10^{-5} (table XXII). These data very clearly demonstrate that the genetic instability of the hybridoma clone during culture passage may cause dramatic changes in the hybridoma product. Therefore, every hybridoma clone should be checked for genetic stability. Instable hybridomas should be cloned specifically to improve genetic stability. This can be done successfully ([72]; see table XXII). Due to low frequency, the occurrence of isotype switch variants does not seem to have any considerable influence on the characteristics of a monoclonal antibody preparation produced by the parenteral hybrid. According to the original finding of Nossal et al. [500], however, they might specifically be selected for improving the effector function of a monoclonal antibody with a given specificity [72, 151, 344, 365, 673].

To summarize, monoclonal antibodies must have sufficient tumor specificity and an utmost affinity as essential requisites for use in tumor immunotherapy. While effector functions of monoclonal antibodies should be low in immunoscintigraphy, radioimmunotherapy and specific chemoimmunotherapy, they are additional prerequisites in tumor immunotherapy. Thus Fab or F(ab')$_2$ fragments of IgG are not suitable in tumor immunotherapy but might be of advantage in those cases where no effector function is required.

Table XXII. Genetic stability of hybridoma BW 494/32 (according to Bosslet et al., 1987 a, b)

Clone	No. of culture passages	Idiotype loss variants	Spontaneous non-producers	Isotype switch variants (IgG$_1$ → IgG$_{2a}$)
Parental	1	0	0	0
	4	16/92	15/92	4/2 x 10^{-5}
Selected for clonal stability	1	0	0	
	15	0	0	
	30	0	0	

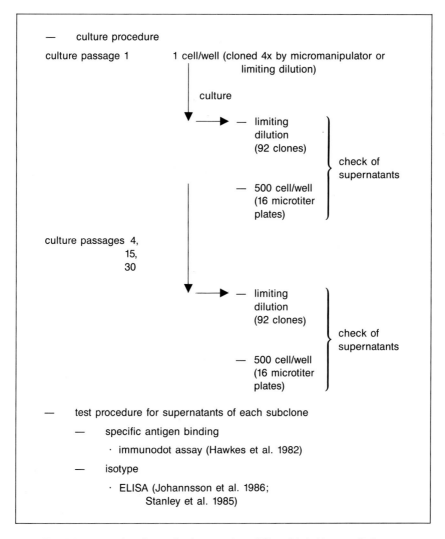

Fig. 4. Test procedure for evaluating genetic stability of hybridoma cell clones.

Quality control for any monoclonal antibody under consideration for use in tumor therapy should guarantee not only the specificity, the effector function and the safety of the antibody preparation, but also its composition, which might be changed by genetic variation of the hybrid.

Radioimmunolocalization

The discovery of TAA and the generation and isolation of antibodies reacting with them has very quickly led to the development of radiolabelled antibodies for the radioimmunolocalization of tumors. In the early 1950s, Pressman and Korngold [540] and Bale et al. [26] already showed that radiolabelled antibodies specifically localized Wagner osteosarcoma and Walker sarcoma of rats. Human choriocarcinomas xenografted in the cheek pouch of hamsters were localized by i.v.-administered radiolabelled antibodies to hCG [548]. Similarly, xenografts of human colon carcinoma were localized by antibodies to CEA [244-246, 543]. This localization was specific, since nonspecific IgG labelled with ^{125}I and simultaneously administered with the ^{131}I-labelled specific antibody showed a lower accumulation at the tumor site. The improvement of the antibody preparation by solid phase immunoabsorption and purification of specific antibodies went hand in hand with better results in tumor localization [425-427, 543]. Experimental evaluation of radiolabelled antibodies was improved by grafting human tumors into T-cell deficient inbred nude mice instead of into the cheek pouch of outbred hamsters [425].

Convincing results in the radioimmunolocalization of human tumors with ^{131}I anti-CEA antibodies were first reported on by Goldenberg et al. [245]. They claimed a diagnostic accuracy of this method in about 85% of all cases. These results were principally confirmed by others [40, 171, 426, 427, 596]. The degree of sensitivity, however, was found to be lower (i.e. about 42%) [426, 427] than first claimed by Goldenberg et al. [245]. The development of monoclonal antibodies significantly accelerated and finally made the broad development of radioimmunolocalization possible. Consequently, monoclonal antibodies were tested in clinical studies very early [194, 426, 427].

Parallel to the improvement of the antibody preparation, the imaging technology was improved to manage the problem of background radioactivity in normal tissues. This background activity is due to the fact that only a small proportion of antibody localizes in the tumor. Furthermore, the

Table XXIII. Technical development and diagnostic efficiency of immunoscintigraphy (according to Baum et al., 1987)

Method	Antibody	Nuclide	Detection limit	Sensitivity
Scanner	polyclonal	^{131}I	> 3.0 cm	43 — 81 %
γ-camera	polyclonal	^{131}I	< 3.0 cm	42 — 85 %
γ-camera + computer	MAb	^{131}I	< 2.0 cm	55 — 91 %
Emission computer-tomography	MAb	^{131}I ^{123}I ^{111}In	< 1.5 cm	90 % > 80 — 90 %

distribution of immunoglobulin in the body is not uniform and this biological fluctuation may mask the signal coming from the activity in the tumor [191, 192]. Images obtained with a gamma camera can be improved by tomographic methods which assess the antibody distribution in three dimensions [42, 191, 192]. This optimization of the imaging technology has increased both the diagnostic sensitivity of the technique and the analytical potency (see table XXIII). For instance, in the case of 16 patients correct results were achieved in about 43% when using planar scintigraphy, whereas with the use of emission computer tomography (ECT) and using the same radiolabelled antibody in the same patients, the degree of correct results could be increased to 94% [427].

Selection of Radionuclides

It is obvious that any radionuclide used for diagnostic radioimmunolocalization must fulfill the following requirements: Its energy must be strong enough and its range of radiation must be wide enough to reach the receiving camera outside the body without difficulties. On the other hand, the accumulated radiation dose must be as low as possible and not harm the patient (see table XXIV). These conditions can only be met by γ-radiators. Thus, a battery of γ-emitting radionuclides has been selected for radioimmunoscintigraphy (see table XXV). Of these, ^{123}I, ^{111}In and ^{99m}Tc all seem to be appropriate. The half-life of these radionuclides is in the range of a few hours or days, and the energy of the γ-radiation between 100 and 200 KeV. Thus the radionuclide can be applied in such a dose as will keep the radiation burden minimal for the patient.

Table XXIV. Conditions for radionuclides conjugated to antibodies applied in vivo

	Immunoscintigraphy	Radiotherapy
Range of radiation	out of the body	restricted to tumor
Radiation dose (accumulated)	as low as possible	as low as possible: outside tumor as high as possible: inside tumor
Type of radiation	γ	β or α
Energy	100—200 KeV	β: $\leq$ 2.5 MeV range: average $\sim$ 4 mm max. < 12 mm α: $\geq$ 2.5 MeV range > 15 μm
Half-life	4—24 h	2—10 days
	depending on pharmacokinetics of MAb	

Table XXV. Nuclides for radioimmunoscintigaphy

Isotope	$T_{1/2}$	Radiation	γ-energy (KeV)
^{123}I	13.2 h	$\beta; \gamma$	159
→ ^{131}I	8.05 days	$\beta; \gamma$	364 (81%)
→ ^{111}In	67.4 h	γ	172, 247
→ ^{99m}Tc	6.0 h	γ	142
^{77}Br	56.0 h	γ	239, 521
^{67}Ga	3.3 days	γ	93, 184
^{67}Cu	62.0 h	$\beta; \gamma$	185 (47%)

For radioimmunolocalization, conjugation of the selected radionuclide to an antibody should take place without damaging its antigen-binding capacity. When released from the antibody the radionuclide should be rapidly and completely excreted from the body in order to prevent any side effects caused by an unspecific accumulation of the radionuclide in certain organs and tissues.

^{131}I was the first suitable radionuclide with which proteins could be radiolabelled [40, 171, 245, 426, 427]. About 0.5–1 mCi are injected to obtain sufficient counts for imaging [191, 192].

^{131}I has certain disadvantages: a considerable amount of free iodine is released in vivo by iodinases and largely excreted through the urinary tract.

However, there is a tendency for free iodine to localize in the thyroid gland and in gastric mucosa in spite of the generally practised blocking with cold iodine or perchlorate [41, 42]. Moreover, ^{131}I emits β-particles and γ-rays of high energy. This means that imaging with conventional cameras is to some extent inefficient [42]. ^{123}I overcomes this problem, having a much more suitable energy for successful imaging [182]. However, its short half-life makes the logistics of supply very difficult.

^{111}In is suitable for radioimmunoscintigraphy, both with respect to energy and to half-life. About 2 mCi has to be injected to obtain sufficient counts for imaging. Being chelated to an antibody [91, 640, 641], the linkage is relatively stable so that the proportion of radionuclide retained in the tumor is higher [41, 42]. The disadvantage, however, is an increased uptake of ^{111}In-labelled antibody mainly in the liver, which makes it difficult to diagnose a tumor-specific localization in that organ. Uptake and enrichment of ^{111}In-labelled antibodies in the liver might be due to pinocytosis and delayed metabolism of ^{111}In-conjugated antibodies by Kupfer cells.

The binding of ^{111}In-transferrin complexes to liver cells via the transferrin receptor has been discussed as an alternative mechanism [307]. Hereby it has been claimed that ^{111}In transchelates from the antibody-bound diethylene-triamine-penta-acetic acid (DPTA) to transferrin, due to the fact that the affinity of DPTA for indium (10^{28} l/mol) is less than that of transferrin (10^{30} l/mol) [733]. However, experimental studies by Ward et al. [724] proved the in vivo stability of ^{111}In-DTPA-MAb conjugates. Thus the biodistribution of indium-labelled murine MAb was not changed by the injection of various chelating agents. In clinical studies only 1–2% of the injected dose has been found to transchelate [307].

^{99m}Tc is very suitable for radioimmunoscintigraphy with respect to its energy, its low cost and ready availability. ^{99m}Tc should be superior to other radionuclides for use in tumor detection via immunoconjugates if localization and blood clearance is rapid enough to take advantage of its short half-life of 6 h. At least 20 mCi has to be injected to obtain sufficient counts for imaging. A new coupling procedure has recently been developed [660] which makes it possible to conjugate ^{99m}Tc to the antibody 'at the bedside' by mixing ^{99m}Tc pertechnetate and a solution of the reduced antibody in the presence of a stannous component. After a short incubation time the preparation is ready for injection. Since ^{99m}Tc can be generated in hospitals just before use, the user is independent of any logistics of an external supply of ^{99m}Tc.

In this regard, the short half-life of ^{99m}Tc is of advantage in immunoscintigraphy. However, just as in the case of ^{111}In, the unspecific uptake of ^{99m}Tc in the liver is increased, possibly via pinocytosis of the immunoconjugate by Kupfer cells.

Methods of Labelling Antibodies with Radionuclides

A successfully labelled antibody is characterized by two main features: completely preserved immunoreactivity and a bond between antibody and radionuclide that is stable under in vivo conditions [11]. To achieve this goal

Table XXVI. Schematic representation of structures on the antibody molecule to bind radionuclides

Anchoring structure	Linker system	Isotopes	Authors
Random labelling			
Tyrosine	electrophil substitution	^{123}I	Wood et al., 1981
	— with oxidants:	^{131}I	Pinto et al., 1977
	chloramine T method	^{125}I	Matzku et al., 1985
	Iodogen® method		Nakamura et al., 1977
	— with lactoperoxidase		von Schenk et al., 1976
	— electrochemically		Rosenberg and Murray, 1979
	diazonium salt technique	^{125}I	Hayes and Goldstein, 1975
		^{111}In	Leung et al., 1978
NH$_2$	DTPA	^{111}In	Atcher et al., 1985
		^{99m}Tc	Chanachai et al., 1985
	N-hydroxy succinimide ester		Najafi et al., 1985
	carbodiimide		Bolton and Hunter, 1973
COOH	carbodiimide	(cross-linking of proteins)	Andres and Schubiger, 1986
Site-specific labelling			
HCO	DTPA (after periodate oxidation of carbohydrates)	^{111}In ^{99m}Tc	Wieland and Fahrmeier, 1970 Borch et al., 1971
SH	DTPA (after reduction)	^{111}In ^{99m}Tc	
		^{99m}Tc	Paik et al., 1985

two techniques have to be differentiated (see table XXVI). Random labelling techniques are characterized by the inability to guide the radionuclide to a defined site on the target antibody, because there are several reactive groups distributed over the molecule, and accessible tyrosines, amino groups and carboxylic acids are among these randomly available groups.

By contrast, site-specific labelling means the selection of a site for radionuclide incorporation which is far away from the antigen-binding domains (idiotypes). Groups for site-specific labelling are the carbohydrate and free sulfhydryl moieties of the antibody [11].

For many years radionuclides of iodine have been used for the simple and quick labelling of antibodies. Of the approximately two dozen well-known iodine isotopes, only the radionuclides ^{123}I and ^{131}I are used for immunoscintigraphical examinations, due to their physical properties and qualities.

Proteins are mainly labelled by means of electrophilic substitution – that is, by replacing hydrogen with iodine in the presence of a mild oxidizing agent. The amino acids tyrosine and histidine are the major targets of this reaction (see fig. 5). The most commonly used methods are the chloramine T and the Iodogen® methods [208], in which the antibody to be labelled is directly transformed by the radioactive iodine.

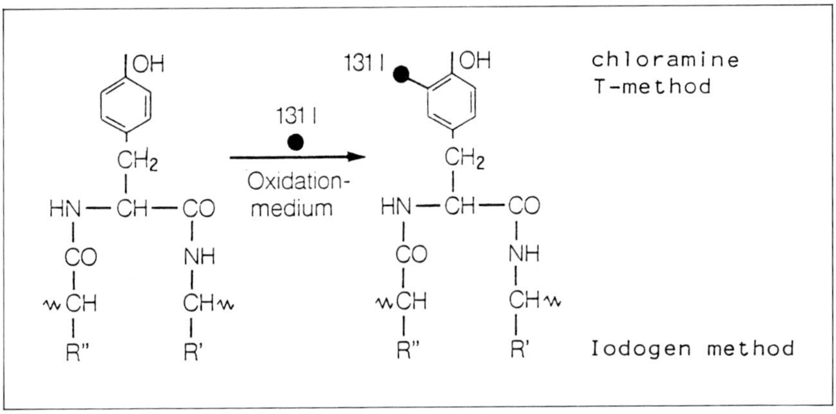

Fig. 5. Conjugation of tyrosine (or histidine) by the replacement of hydrogen by iodine in the presence of mild oxydants (electrophil substitution). Advantages: simple and quick procedure; disadvantages: in vivo instability by enzymatic cleavage (deiodases) in RES; high accumulation in the thyroid gland.

We prefer a modified Iodogen method [567] in which the oxidizing agent is dissolved in acetone and then added to the aqueous solution of antibody and radioactive iodide. The resulting reaction takes place in a homogeneous phase and supplies reproducible yields of 90–95%, whereby a reaction time of 10 min suffices.

The advantage of this method is in the simplicity and speed of the labelling procedure. For this reason iodine labelling must be regarded as the first-choice method for the new testing of an antibody. It provides a rapid, preliminary survey regarding the accumulation in the target organ, the kinetic behavior of the protein and the distribution pattern in other organs. A disadvantage, however, is the lack of in vivo stability of these test materials. Since in vivo deiodination is undesirable, coupling reagents have been developed in which the radioiodine is bound to a non-activated aromatic ring and thereafter binds covalently via a functional group with reactive substituents in the antibody.

Iodine-labelled proteins can be obtained by halogen replacement over an intermediate grade of a metal organic compound [323, 631] (fig. 6). These proteins have a low in vivo deiodination and therefore a low accumulation in the thyroid gland.

Metallic elements cannot take the reaction course of an electrophilic aromatic substitution. In order to label monoclonal antibodies with metallic, radioactive nuclides, bifunctional chelate components are used. On the one hand these substances have the attribute of binding different metals to a complex, and on the other hand they can form covalent links

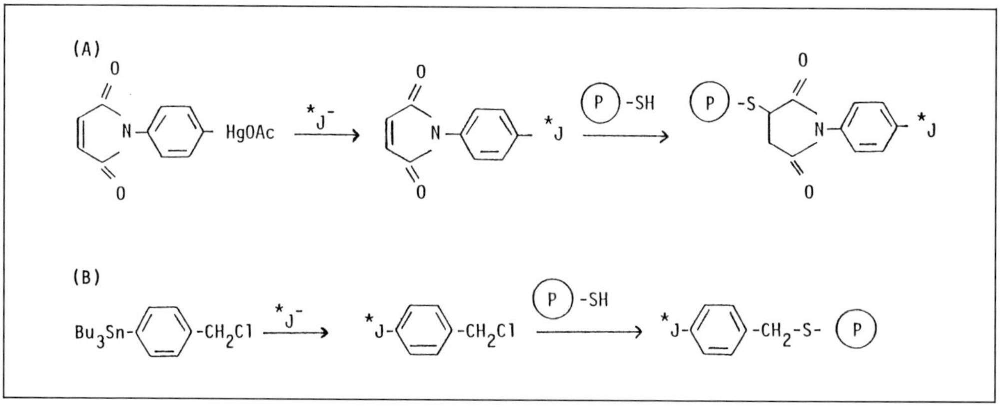

Fig. 6. Iodine labelling of monoclonal antibodies via metal-organic compounds.

to certain amino acids of proteins. Figure 7a,b shows a compilation of bifunctional complex components for radioactive labelling of monoclonal antibodies [81, 138, 450, 653]. Apart from the structures shown in this figure, there are further complex components coupled to proteins: cryptands [223b], deferoxamin [476] and metallothionein [682]. However, the most commonly used reagent is the bicyclic anhydride of diethylene-triamino-penta-acetic acid (DTPA). The reason for the popularity of just this bifunctional complex component probably lies in its simple synthesis, good storage capacity, high reactivity and altogether easy operating techniques.

The linking of the chelate component with the monoclonal antibody or general protein is of particular significance. On the one hand it should be effective – that is, produce a high reaction yield – and on the other hand it must on no account have a negative influence on the biological properties of the protein. At first glance these two demands seem to run contrary to each other and thus appear unrealizable, especially when taking into consideration that almost every known coupling reaction does not run according to a plan [7]. The linking does not take place with a particular amino acid at a particularly chosen spot of the protein but rather by random distribution over the whole of the protein of the amino acids which are capable of reacting. This means that one must reckon with irreversible damage to the protein, even by small, bound parts of the complex component. Knowing this, it is important that the coupling reaction for every single antibody is carefully optimized and controlled. The following parameters are of particular importance:
- protein concentration,
- molar excess on the reagent.

The optimal course of the coupling reaction cannot really be checked until the radioactive labelling and measuring of the biological property in question have been carried out. In the case of monoclonal antibodies, this is the in vitro or in vivo analysis of the reaction capability with the corresponding specific antigen. In addition, good possibilities of optimizing the coupling reaction are offered by gel filtration in high performance liquid chromatography. This method enables a quick and simple recognition of the degree to which the retention behavior of the antibody is changed by the reaction terms and/or the degree to which the undesired high moleculed protein parts appear. It is this side reaction which plays an important role in the use of the bicyclic anhydride of the DTPA [306, 492, 518, 519].

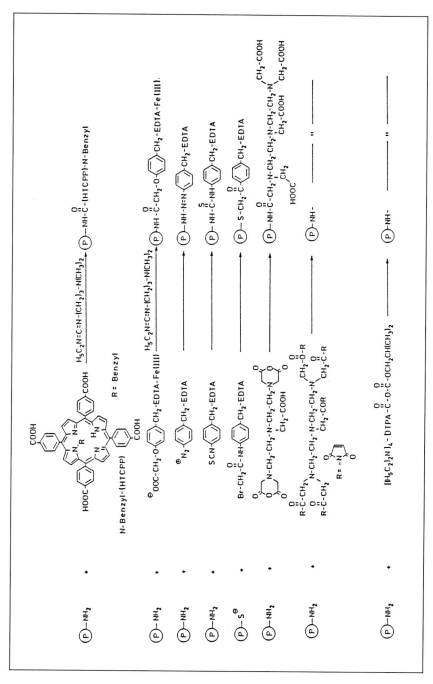

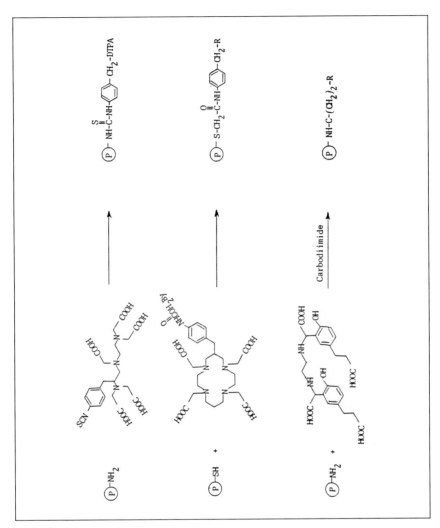

Fig. 7. a, b Labelling of antibodies with metallic nuclides by bifunctional chelating compounds.

Our own experiences with the coupling reaction of bicyclic DTPA anhydride with an antibody fragment have shown that the immune reactivity of the monoclonal antibody fragment is only slightly decreased when an optimal choice of reaction conditions are given, and that at the same time a labelling yield with ^{111}In of more than 90% can be achieved.

The exact analysis of the amount of complex component actually bound to the antibody is still a great problem. Meares et al. [449, 450] worked out a method for the quantitative determination of linked DTPA, which used cobalt 57 as radioactive nuclide. However, when we tested this method the results showed a relatively high variance. The use of the ^{14}C-labelled reagent [723] seems to offer far more exact results. In addition, this method has the advantage of optimizing the coupling reaction more than was possible up to now.

Special methods have been developed to label monoclonal antibodies with ^{99m}Tc, which is particularly suited to diagnosis in nuclear medicine.

By means of a 'pretinning method' [566] it is possible to label F(ab')$_2$ fragments of antibodies directly with ^{99m}Tc. However, the result is always a mixture of the labelled F(ab')$_2$ and Fab fragments which must be purified, i.e. separated from impurities.

Free thiol groups in the antibody are labelled with ^{99m}Tc in a labelling procedure which is simple and feasible for clinical routine examinations [660]. First the intact antibody is reduced with a thiol (e.g. 2-mercaptoethanol), and then in a purified form lyophilized together with phosphate puffer. The ^{99m}Tc labelling is carried out by taking the reduced antibody and adding a small amount of tin-II-containing labelling unit normally used for skeleton or liver scintigraphy, dissolved for this purpose in an isotonic NaCl solution. After the addition of a ^{99m}Tc pertechnetate solution, the result is a quantitatively labelled antibody which can be directly applied without further purification.

P-carboxyethylphenylglyoxal-di-(N-methylthiosemicarbazone) was synthesized as a special bifunctional chelate agent for the ^{99m}Tc-labelling of antibodies [12]. This compound (fig. 8) contains di-(N-methylthiosemicarbazone) for the complexing of technetium and an aralkyl-carboxylate residue for the coupling to the antibody.

The very stable technetium-diamide-dimercaptide complex [213] (^{99m}Tc-N$_2$S$_2$) which is to a large extent excreted renally, can be used for antibody labelling with technetium over a further functional group (e.g. an activated ester group). The labelling of antibodies which have been conjugated with the ligand before labelling produces yields of 90%, while the conjugation of the labelled N$_2$S$_2$ chelate with the antibody only produces yields of 50-70%.

The labelling degree of antibodies should not be higher than 1 to minimize radiological damage of the protein due to the decay of the radionuclides causing local doses of Auger cascades [417] acting on the car-

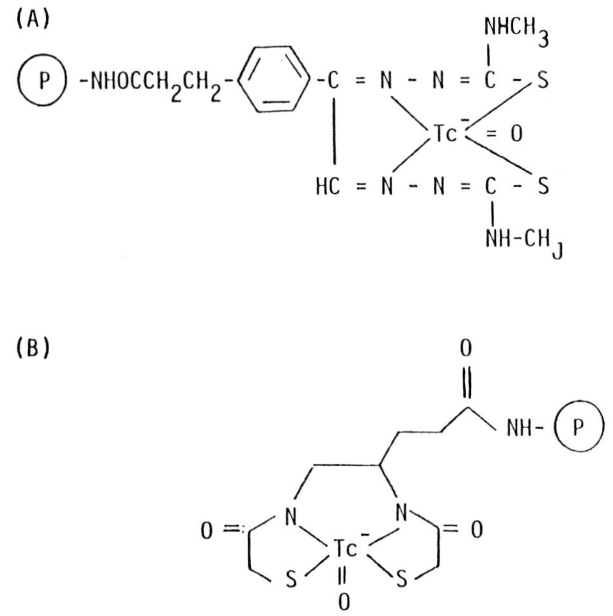

Fig. 8. Bifunctional chelates for ^{99m}Tc-labelling of antibodies.

rier antibodies. Moreover, conversion electrons and γ-radiation by ^{131}I may destroy antibodies in the vicinity. Andres and Schubiger [11] claim that radiolysis by radioactive iodine may at least partly be overcome by avoiding a high antibody concentration and/or by the addition of radioprotective materials – for instance, albumin (20 mg/ml) – to the solvents. The radioprotective activity of albumin, however, is questionable.

Analysis of Immunoreactivity

Labelled monoclonal antibodies used for immunoscintigraphy or immunotherapy must not lose their quality of forming an immune complex with the respective antigen while being processed. The exact amount of radioactivity fixed to the antibodies must be determined for every single preparation. These antibodies are able to build an immune complex with its immune reactivity. The amount of active antibodies forming immune complexes is noted in percent of the whole existing activity and described as the immune reactivity of a preparation. If the value is not 100% (with the exception of methodical variations of the test), then part of the activity is present in an undesired form such as labelled but completely defective antibodies or primarily unreactive antibodies. Further undesired forms are free activity or the existence of other labelled components in the preparation apart from antibodies.

The analysis is carried out with an equilibrium assay in antigen access. Tumor cells are used as carriers of the respective antigens.

While the number of antibodies within one test remains the same, the cell concentration is increased until every reactive (i.e. undamaged) antibody has the possibility to bind itself to a cell. After centrifugation the cell-bound parts and the activity in the supernatant are divided and the activity shares are determined.

The cells used originate from an antigen-positive human tumor and are either multiplied by cell culture or nude mouse (nude rat) xenotransplantate. The fixation method (formaldehyde fixation, cryopreservation) depends on the stability of the defined epitope.

After the incubation of cells and antibodies the results are quantified. The cell-bound activity shares are calculated (in percent of displayed activity) and plotted as a function of the cell mass. In the case of a large cell mass (that is, in the case of antigen access) the curve should reach a plateau (see fig. 9). The resulting percentage (plateau value) shows the immune reactivity of the preparation. For a more exact determination of this value, the reciprocal values of the cell-bound activity are plotted in relation to the reciprocal cell mass. In each case this results in a straight line where the ordinate represents the extrapolated reciprocal value of the immune reactivity. In this manner the immune reactivity can be determined even if the plateau has not yet been completely reached (see fig. 10).

This kind of evaluation, known as the Lineweaver-Burk method, was first described by Lindmo et al. [415] as being applicable in this situation.

A modification of this method makes it possible to determine the binding constants of MAb and the number of epitopes on the cell by means of a Scatchard plot evaluation. In this case the antigen (cell) number remains constant and the antibody number is varied. For the calculation of the binding constant, information regarding the immune reactivity and the unspecific binding (determined with the help of an antigen-negative cell line and/or by an iteration method within the test) is necessary.

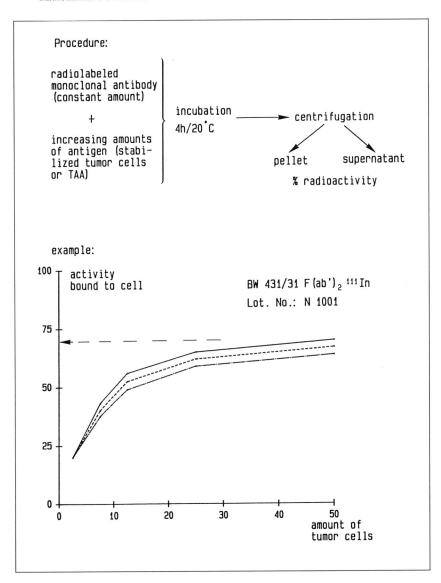

Fig. 9. Immunoreactivity of radiolabelled monoclonal antibodies (relative and objective measurements).

A further possibility of checking the 'immune reactivity' is the competitive binding assay. This test [307] is, however, more complicated and not as safely quantifiable. The antibody to be tested is separated from its

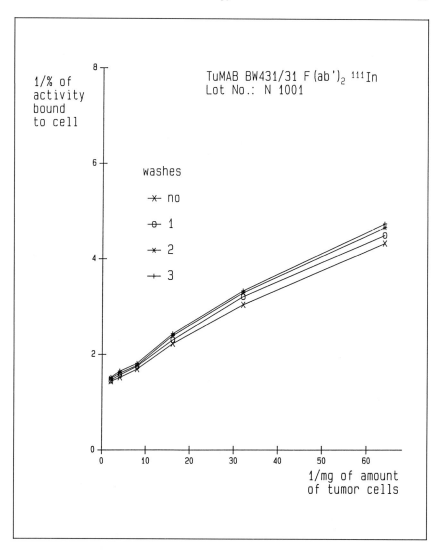

Fig. 10. Immunoreactivity of radiolabelled monoclonal antibodies (reciprocal values are plotted).

antigen by the addition of increasing amounts of the same antibody with different (or no) labelling. The same procedure is carried out for a reference MAb. The comparison of the binding curves gives information about the degree of possible damage due to labelling.

Experimental Tumor Imaging

Experimental studies for immunoscintigraphy of tumors with radiolabelled monoclonal antibodies have to be done in in vivo models. Either experimental tumors of mice and rats or human tumors xenografted into immunodeprived mice are used [579, 580]. In the case of experimental tumors, fundamental aspects of radioimmunoscintigraphy with labelled monoclonal antibodies specific for that selected murine or rat tumor can be elaborated. Such information can be extrapolated for the clinical situation. If, however, the qualification for immunoscintigraphy of a radio-labelled monoclonal antibody specific for a given human TAA has to be tested preclinically, the xenograft system is the only available experimental model in providing information pertinent to the human situation. Human tumors are either transplanted into the subcutis, the peritoneal cavity, the lung, the gut, the liver, the spleen or into the cerebral hemisphere [406, 609, 627]. The limitations of the xenograft model, however, must be regarded critically.

Thus it is known that the amount and the type of TAA being expressed in xenografts varies quantitatively [173, 289] as well as qualitatively [739], compared to the corresponding naturally occurring human tumors (see p. 5ff.). Moreover, the monoclonal antibodies specific for human TAA are truly 'tumor specific' in mice or rats xenografted with human tumor tissue. By contrast, in man there can be at least a slight positive reaction with some normal tissues [182]. In addition, Fc-mediated functions and Fc-dependent distribution of murine MAb in nude mice are certainly different from those in man. Apart from immunological reactivity, other factors relating to tumor vascularity, lymphatic drainage and fibrotic reactions and encapsulation may influence antibody accumulation at the tumor site [579, 580] and thus cause significant differences between the original and the xenotransplanted human tumor. Such differences in the expression and accessibility for monoclonal antibodies of TAA may explain the contradictory results in radioimmunoscintigraphy of xenografted and original human tumors of similar type [273, 579, 580].

An additional reason for discrepancies between experimental and clinical radioimmunoscintigraphy is the size of the tumor compared to body weight and blood volume. The relative mass of a tumor (0.1–4.0 g) xenografted into and growing in a nude mouse (15–30 g) is in the range of 0.3%–30%. By contrast, the relative volume of a tumor (2–10 g) in a patient (75 kg) is in the range of 0.003–0.01% (see table XXVII). Conse-

Table XXVII. Systemic radioimmunotherapy — dosimetric aspects of radionuclide-labelled antibodies in mouse and man

	Antibody/dose applied	Relative tumor volume $\left(\dfrac{TW}{BW} \times 100\right)$	Relative amount of specific antibody localization (% antibody/g tumor)
Mouse (nu/nu; colon carcinoma, s.c. transplant)	BW 431/31 - ^{111}In (Fab')$_2$ (2 μg/mouse)	0.7 — 27.0 (TW = 0.1 — 4.0 g)	20.4 (10 — 20)
	BW 431/26 (IgG) ^{131}I ^{111}In ^{99m}Tc $\Big\}$ (2 μg/mouse)	0.7 — 27.0 (TW = 0.1 — 4.0 g)	10 — 30
Man (colon, breast, ovarian carcinoma)	HMFG 2 - ^{131}I* 5 — 10 μg/i.v. (5 — 8 mCi/mg protein)	0.003 — 0.01 (TW = 2 — 10 g)	0.015 (0.002 — 0.032)
(colon carcinoma)	BW 431/31 - ^{111}In (Fab')$_2$ 1 mg i.v. (1 — 2 mCi/mg protein)	0.003 — 0.01 (TW = 2 — 10 g)	~ 0.01

* Data from Epenetos et al., 1986
TW = Tumor weight
BW = Body weight

quently, the absolute amount of antibody localizing in tumors of patients is generally much less than that found in experimental systems. Moreover, the larger blood volume in the patient may lead to a lower initial concentration and a slower turnover of localizing antibody in the blood pool. This in turn would reduce the net uptake in the tumor [579, 580]. Indeed, pharmacokinetic studies by Harwood et al. [273] have shown that the rate of antibody accumulation in tumor xenografts is proportional to its concentration in the blood pool.

Most of the kinetic data, however, have to be regarded with caution if iodine-labelled monoclonal antibodies or fragments of them have been used. These iodine-labelled monoclonal antibodies are sensitive to deiodinases occurring in the body [550, 754]. Free iodine may simulate a distribution of the antibody which does not conform to reality [182, 307, 361].

In mice, the amount of radiolabelled antibody localizing in the xenograft (melanoma, colon carcinoma and various other tumors) was found to be between 0.5% and 50% of the injected dose, irrespective of the radionuclide being selected [22, 235–237, 273, 429, 430, 439, 579, 580, 639–641]. In our own experiments with stably ^{111}In-labelled F(ab)$_2$ fragments or ^{131}I, ^{111}In or ^{99m}Tc-labelled intact monoclonal antibodies, about 10–30% of the injected dose localized in the tumor within about 1 day (see tables XXVIII, XXIX). By contrast, tumor uptake of specific antibodies in patients ranged from 0.002% to 0.03% (mean 0.015%) of total injected dose per g of tumor 1 day after i.v. application (see table XXVII).

To overcome the higher 'dilution factor' of monoclonal antibodies in patients, much higher amounts of mouse immunoglobulins (about

Table XXVIII. Organ distribution of ^{111}In MAb BW 431/31 F(ab')$_2$ (3 nude mice each)

Sacrificed post injection		3 h	7 h	17 h	27 h	48 h	72 h
Liver	(%/g)	8.9	9.2	9.6	6.8	6.7	4.8
Lung	(%/g)	8.7	6.5	3.7	1.9	1.1	2.1
Spleen	(%/g)	4.2	3.9	5.0	4.1	4.3	3.3
Muscle	(%/g)	0.6	0.6	0.7	0.5	0.3	0.4
Tumor	(%/g)	5.3	7.7	20.4	16.0	15.0	10.1
Blood	(%/ml)	17.6	12.4	4.8	2.1	0.5	0.3
Kidneys	(%)	13.8	18.8	25.4	20.3	15.7	15.3
Gut	(%)	6.3	8.1	5.9	4.5	3.7	3.3
Carcass	(%)	50.1	44.8	43.0	32.4	9.8	13.8

Table XXIX. Estimation of time dependency of tumor localization of MAb BW 431/26 in nude mice

Time:	5—7 h			17 h			27—30 h		
Radionuclide:	^{131}I	^{111}In	^{99m}Tc	^{131}I	^{111}In	^{99m}Tc	^{131}I	^{111}In	^{99m}Tc
Tumor (% antibody/ g tumor):	4.8	8.2	8.6	14.9	9.8	11.0	9.4	12.5	14.9

20–50 mg per patient) would be needed to achieve comparable tumor uptake [184]. This topic has not been explored much because the total dose of radiation which can be given is limited to doses acceptable for safety in a diagnostic procedure [42]. Data of Pimm and Baldwin [530] show that increasing doses of antibody in an animal tumor model result in correspondingly increased amounts of antibody in the tumor. This finding, however, contradicts our own experience.

In patients, the application of a high dose, i.e. 20 mg, of monoclonal antibodies could achieve prolongation of t1/2 in the blood, but it is not yet clear whether there is improvement in the tumor: non-tumor ratios [684]. It has been stated that the amount of monoclonal antibody localizing in the tumor correlates with the concentration in the blood pool just as the amount of monoclonal antibody localizing in normal organs [273]. The only difference seems to be the mode of clearance of the antibody. In normal tissue the antibody concentration starts to decrease after about 4h, whereas in tumor tissue it is maintained much longer [133–136]. This is thought to be caused by TAA-antibody immune complexes, which form a lattice network in the extravascular space at the tumor site and prevent clearance of the antibody by lymphatic drainage [273, 579, 580].

However, this direct correlation was found to depend on tumor size, with the smaller tumor accumulating significantly more antibody (60 μg/g tumor) than the larger one (5.2 μg/g tumor) [580]. A reason could be the reduced perfusion of greater tumors and/or resulting necrosis. Moreover, increased doses of radiolabelled antibody also caused an increase in the levels of radioactivity associated with all normal tissues studied (reducing the ratio of tumor to normal tissue to about 50%) and diminished the cumulative localization of the radioactivity in the tumor relative to normal tissue clearance [579, 580]. These studies point to the necessity of accelerating clearance of circulatory radiolabelled antibody. At least experimentally, this may be achieved by using a second 'clearing' antibody for the removal of non-tumor-bound intact anti-tumor antibodies after localization at the tumor site has taken place [30, 41, 79, 610]. Another possibility might be the use of antibody fragments as potential vehicles for radionuclides [579, 580].

In the case of stably labelled F(ab')$_2$ fragments, however, it could be shown that their clearance out of the body is not as different from the clearance of intact IgG as may be assumed from the data gained with iodine-labelled fragments [182, 307, 361, 667]. Thus the accelerated clearance of iodine-labelled F(ab)$_2$ fragments is simulated by free iodine, cleaved from the iodine-labelled antibody by deiodinases.

Results are contradictory regarding the question of whether an increase in tumor size may lead to a higher accumulation of the antibody. In experimental systems, either no relationship [283, 284] or a linear relationship has been found – by both Baldwin and Pimm [24] and Mann et al. [439] – whereas Moshakis et al. [473, 474], Hagan et al. [260] and Philben et al. [526] revealed an inverse relationship between tumor size and uptake of specific monoclonal antibodies. This inverse relationship was neither tumor specific (colon, melanoma, lymphoma) nor radiolabel specific (^{125}I, ^{111}In) [260]. However, a clear-cut correlation between the amount of TAA being produced by the respective xenograft and the amount of antibody localization at the tumor site could be found [531–533]. When the monoclonal antibody was not specific for the tumor, a statistically significant correlation to size did not occur. The contradicting results regarding the degree of tumor localization and tumor size may thus be caused by differences in the amount of the TAA antigen, produced and/or excreted by the respective tumor xenografts [531–533]. It could also be caused by different affinities [275] or different degrees of cross-reactivity [139, 283, 284, 544–546, 719] of the antibody and by the variation of its in vivo survival and kinetics of tumor localization [89, 283, 284].

To increase the concentration of antibody reaching the tumor, the regional administration of the radiolabelled antibody was tried as an alternative to systemic application via i.v. injection. Administration of the antibody via lymphatics showed a greater proportion reaching the tumor in local lymph nodes than when the antibody was given intravenously [688, 689, 731].

Numerous attempts have been made to improve radioimmunoscintigraphy by increasing the amount of antibody localizing at the tumor site after systemic application. This amount seems to depend on the following parameters:

As has already been discussed (see p. 5 ff. and p. 20 ff.), the quality of the antibody is critical. This includes the selection of a specificity recognizing a TAA, which is then exposed in sufficient amounts by most if not all of the cells of the respective tumor. This TAA should be a membrane constituent, not shedded or secreted into intercellular space and blood. This, however, is an idealistic postulate. Usually, the TAA (to which the selected monoclonal antibody binds specifically) is more or less secreted by the tumor cell and present in the blood, at least in minor amounts. Despite these circumstances, successful tumor imaging seems possible [544-546]. The reason for this might be that the dose of the antibody administered exceeds the blocking capacity of free antigen or that TAA-antibody immune complexes in the blood may retain a free antigen-combining site capable of reacting at the tumor site. Nevertheless, the selection of a monoclonal antibody, mainly binding to an epitope exposed on a TAA like CEA, located in the cell membrane but minimally on the same TAA in solution, as has already been found by Bosslet et al. [74], is a significant improvement which approaches the above-mentioned idealistic postulate.

Taking into account the heterogeneity and the quantitative limitation of TAA expression in tumors, the proposal was made to improve radioimmunoscintigraphy by simultaneously applying several different monoclonal antibodies instead of just one. In preliminary experimental investigations [220, 482, 483] as well as in preliminary clinical studies [119] such mixtures of different non-cross-reacting monoclonal antibodies were superior to the application of a single preparation. Future work should be done to prove and to confirm those data.

Another attempt at improving radioimmunolocalization is the use of $F(ab')_2$ or Fab fragments instead of intact IgG. This reduces increased binding of the monoclonal antibody via its Fc region to the reticuloendothelial system, including lymphatics, liver and spleen (see p. 28 ff.). These organs are known to be the major site of immunoglobulin catabolisms [215]. A reduction of the catabolism of the antibody could lead to an increased antibody uptake in the tumor site [184]. Indeed, an improved tumor-to-non-tumor ratio of Fab and of $F(ab')_2$ fragments has been reported on in experimental studies where mouse immunoglobulins and their fragments were compared simultaneously [89, 134, 294, 720]. The same was found in clinical studies [98, 343, 429, 430].

The higher tumor-to-normal-tissue ratio was interpreted as being due to a more rapid clearance of fragments than intact antibody from the normal tissue and from the body.

In addition, the absolute amount of antibody which bound to the tumor was lower for Fab and $F(ab')_2$ antibodies than for intact IgG [89, 235-237, 294, 720]; moreover, the $F(ab')_2$ radioactivity faded from the tumor faster than the radioactivity of intact antibody. Again, this was interpreted as showing either there is a higher dissociation and loss of $F(ab')_2$ from the tumor or that a more rapid catabolism takes place at the tumor site [720].

All these experiments, however, were carried out with iodine-labelled antibodies and antibody fragments. As has already been pointed out, the data of Khaw et al. [361], Epenetos et al. [182], Hnatowich et al. [307] and Steinstraesser et al. [667] strongly indicate that higher susceptibility of iodinated immunoglobulin fragments to ubiquitous dehalogenase enzymes [550, 754] may be responsible for differences in the loss of label between the intact antibody and Fab or F(ab')$_2$ fragments. This difference in the amount of free iodine may simulate an increased clearance and degradation of Fab or F(ab')$_2$ fragments.

If dehalogenase-resistant radiolabels are used, such as ^{111}In, a higher tumor uptake of the antibody [49, 191, 192] can be achieved even with F(ab')$_2$ fragments, but at the same time a higher uptake of the antibody by normal liver and kidneys may occur [49]. It is thus obvious that the data obtained from iodinated monoclonal antibody preparations clearly differ from the data obtained when using 'stably' labelled antibody. It is also clear that in the case of these preparations which are easier to deiodinize (e.g. antibody fragments) there is a particularly large discrepancy between the radionuclide distribution and the protein distribution. Thus preclinical investigation is of imminent importance in finding the optimum labelling procedure.

We have elaborated the distribution of the anti-CEA antibody MAb BW 431/31-F(ab')$_2$ labelled with ^{111}In in the xenograft model (see table XXVIII). Here the highest tumor value (20% of the injected dose/g tumor) was reached after 17 h. This value lay in the same range as the value achieved with intact immunoglobulins linked with different radionuclides (see table XXIX).

In spite of the fact that even F(ab)$_2$ fragments failed to have a higher clearance rate than intact immunoglobulin, there is still a chance for Fab fragments or fragments even smaller than Fab fragments to reach this goal. Such fragments should have the unimpaired potency of antigen binding but should be superior to all other antibody preparations with respect to penetration of tissues and elimination rate. It should be possible to construct such fragments using genetic engineering methods. This would also enable us to modify the amino acid sequence of the constant part of the antibody fragment with a view to reducing any unforeseen unspecific binding to tissues and optimizing the linkage of a radionuclide.

Clinical Data

A considerable number of investigators have already tested monoclonal antibodies for immunoscintigraphy of tumors in patients. The detected

Table XXX. Radioimmunoscintigraphy — clinical results

Author	Antigen/antibody	Isotype	Radio-nuclide	No. of patients	Correct results (%)
Gastrointestinal tumors					
Mach et al. (1981)	CEA	F (ab')$_2$; IgG$_1$	^{131}I	28	50
Mach et al. (1981)	CEA	F (ab')$_2$	^{131}I	14	90
Delaloye et al. (1986)	CEA	Fab; F (ab')$_2$; IgG$_1$	^{131}I	44	86
Mach et al. (1983)	17—1A	F (ab')$_2$/ IgG$_{2a}$	^{131}I	63	54
Moldofsky et al. (1983)	17—1A	F (ab')$_2$/ IgG$_{2a}$	^{131}I	32	69
Chatal et al. (1985)	17—1A	F (ab')$_2$/ IgG$_{2a}$	^{131}I	46	59
Smedley et al. (1983)	CEA	IgG$_{2a}$	^{131}I	16	81
Chatal et al., (1985)	Ca 19—9	F (ab')$_2$ IgG$_1$	^{131}I	29	66
Chatal et al. (1985)	Ca 19—9	F (ab')$_2$	^{131}I	29	66
	17—1A	F (ab')$_2$		13	77
Farrands et al. (1982)	791T/36	IgG$_{3b}$	^{131}I	11	82
Armitage et al. (1985)	791T/36	IgG$_{2b}$	^{111}In	16	75
Armitage et al. (1985)	791T/36	IgG$_{2b}$	^{131}I	50	34
Baum et al. (1986)	CEA	F (ab')$_2$	^{131}I	75	91
Baum et al. (1987)	Ca 19—9	F (ab')$_2$/ IgG$_1$	^{131}I	151	86
Montz et al. (1986/87)	Ca 19—9 + CEA	F (ab')$_2$	^{131}I	34	79
Chatal et al. (1986)	CEA + Ca 19—9	F (ab')$_2$; IgG$_1$	^{131}I	64	75
Chatal et al. (1986)	Ca 19—9	F (ab')$_2$ IgG$_1$	^{131}I	46	72
Perkins et al. (1986)	CEA	IgG$_{2a}$	^{111}In	17	88
Allum et al. (1986)	CEA	IgG$_1$	^{131}I	35	66
Buraggi et al. (1985)	CEA	F (ab')$_2$; Fab	^{111}In ^{131}I ^{99m}Tc	25	80
Riva et al. (1985)	CEA	F (ab')$_2$; Fab	^{99m}Tc	100 i.v. 33 i.p.	78
Scheidhauer et al. (1986)	CEA + Ca 19—9	F (ab')$_2$; IgG$_1$	^{131}I	16	94
Chatal (1987)	CEA + Ca 19—9	F (ab')$_2$; IgG$_1$	^{131}I	23	74

Table XXX. (continued)

Author	Antigen/antibody	Isotype	Radio-nuclide	No. of patients	Correct results (%)
Baum et al. (1987)	CEA + Ca 19—9	F (ab')$_2$; IgG$_1$	^{131}I	54	85
Gynecological tumors					
Epenetos et al. (1982)	HMFG$_{1/2}$	IgG$_1$	^{131}I	39	62
Williams et al. (1984)	791T/36	IgG$_{2b}$	^{131}I	17	47
Symonds et al. (1985)	791T/36	IgG$_{2b}$	^{131}I	12	91
Perkins et al. (1985)	791T/36	IgG$_{2b}$	^{131}I	18	89
Epenetos et al. (1985)	Plac. AP	IgG$_1$	^{111}In	15	66
Granowska et al. (1984)	HMFG$_2$	IgG$_1$	^{123}I	40	88
Granowska et al. (1986)	HMFG$_2$	IgG$_1$	^{123}I	20	95
Pateisky et al. (1985)	HMFG$_2$	IgG$_1$	^{123}I	18	89
Jackson et al. (1985)	Plac. AP	IgG$_{2b}$	^{123}I	13	76
Thompson et al. (1984)	3E1—2	IgM	^{131}I	9	100
Baum et al. (1986)	Ca 125	F (ab')$_2$	^{131}I	71	82
Bourguet et al. (1986)	Ca 125	F (ab')$_2$	^{131}I	28	86
Chatal et al. (1986)	Ca 125 + Ca 19—9	F (ab')$_2$	^{131}I	24	67
Chatal et al. (1986)	Ca 125	F (ab')$_2$	^{131}I	94	82
Baum et al. (1987)	Ca 125	F (ab')$_2$	^{131}I	28	64
Doherty et al. (1986)	Ca 125	F (ab')$_2$	^{111}In	8	75
Rainsbury et al. (1984)	M 8	IgG	^{111}In	17	76
Chatal et al. (1987)	β HCG	F (ab')$_2$	^{131}I	5	80
Fridrich et al. (1986)	breast cancer antigen	F (ab')$_2$	^{123}I	14	77
Melanoma					
Larson et al. (1985)	p97	IgG$_1$; IgG$_{2a}$	^{131}I	25	88
Larson et al. (1985)	HMWA/48.7	Fab	^{131}I	23	74
Halpern et al. (1985)	p97/96.5	IgG$_{2a}$	^{111}In	79	61
Buraggi et al. (1984)	HMWA/225.28S	IgG$_{2a}$; F (ab')$_2$	^{123}I ^{111}In ^{99m}Tc	38	68
Murray et al. (1985)	HMWA/225.28S	IgG$_{2a}$; F (ab')$_2$	^{111}In	100	50
Buraggi et al. (1985)	HMWA/225.28S	IgG$_{2a}$; F (ab')$_2$	^{131}I	18	56
Buraggi et al. (1986)	HMWA/225.28S	IgG$_{2a}$; F (ab')$_2$	^{123}I ^{131}I ^{111}In ^{99m}Tc	101	73

Table XXX. (continued)

Author	Antigen/antibody	Isotype	Radio-nuclide	No. of patients	Correct results (%)
Buraggi et al. (1986)	HMWA/225.28S	IgG$_{2a}$; F(ab')$_2$	^{99m}Tc	29	90
Buraggi et al. (1986)	HMWA/225.28S	IgG$_{2a}$; F(ab')$_2$	^{99m}Tc	40	70
Scheidhauer et al. (1986)	HMWA/225.28S	IgG$_{2a}$; F(ab')$_2$	^{99m}TC	9	89
Scheidhauer et al. (1986)	HMWA/225.28S	IgG$_{2a}$; F(ab')$_2$	^{99m}Tc	34 / 34	62 / 85
Riva et al. (1985)	HMWA/225.28S	IgG$_{2a}$; F(ab')$_2$	^{99m}Tc	115	75
Masi et al. (1985)	HMWA/225.28S	IgG$_{2a}$	^{111}In	46	59
Carrasquillo et al. (1984)	HMWA/9227	IgG$_{2a}$	^{111}In	22	86
Neumann et al. (1985)	p97/96.5	IgG$_{2a}$	^{111}In	45	67
Murray et al. (1985)	GP240/ZMW-018	IgG	^{111}In	46	43

tumors were mostly larger than 1 cm in diameter. Baum et al. [38] collected most of the clinical data in gastrointestinal, gynecological tumors (mainly ovarian carcinoma, less in mammary carcinoma) and melanoma. Summarized information on these clinical studies is given in table XXX. When all these studies are analyzed, it is astonishing to recognize that the ratio of correct results varies in the range of 40–90% (see table XXXI), irrespective of whether F(ab')$_2$ or Fab fragments or which IgG subclass were used, which radionuclide was linked to the antibody and which technique for radioimaging was applied. This might in part be due to the different antibody specificities which were used. However, as outlined in table XXX, the number of antibody specificities tested clinically was quite restricted. Another reason for the variation might be differences in the quality of the antibody preparations after linkage to the respective radionuclide and differences in optimal adaptation of the technical device for radioimaging. Whatever the cause of the variations, they seem to neutralize or obliterate all the improvements which have been elaborated preclinically and clinically confirmed in selected, comparative studies – e.g. the superiority of ^{131}I-labelled (F(ab')$_2$ fragment over identically labelled intact IgG in the imaging of gastrointestinal tumors [465] or the development achieved in the imaging technical device.

Table XXXI. Immunoscintigraphy of tumors (based on data collected by Baum et al., 1987, extended)

	Gastrointestinal tumors		Gynecological tumors		Melanoma	
	no. of studies	correct results (%)	no. of studies	correct results (%)	no. of studies	correct results (%)
Studies						
Retrospective	22		12		15	
Prospective	3		5		2	
No. of monoclonal antibodies	10		10		10	
F (ab')$_2$	17	75 (50—91)	8	76 (64—86)	12	73 (50—90)
Fab	1	86	—		2	74/88
IgG	7	81 (66—94)	9	68 (47—95)	5	63 (43—86)
IgG$_1$	2	80 (66—94)	3	83 (66—95)	1	61
IgG$_{2a}$	2	81/88	—		3	71 (59—86)
IgG$_{2b}$	3	75 (68—82)	4	76 (47—91)	—	
Radionuclides						
^{131}I	20	75 (50—93)	9	80 (47—100)	4	73 (56—88)
^{111}In	4	82 (75—88)	3	72 (66—76)	8	64 (43—86)
^{123}I	1	86	6	82 (62—95)	1	68
^{99m}Tc	1	88	0		7	78 (62—90)
Technique						
planar γ-camera	14	79 (50—91)	11	84 (47—100)	6	71 (56—88)
SPECT (ECT)	9	79 (54—94)	8	74 (62—82)	11	70 (43—90)
Total no. of patients	1,035		505		804	
Correct results	795	77	397	79	537	67

Facing this problem, we improved and perfected the technique of radioimmunoscintigraphy, both with respect to the selection and quality of the antibody as well as to the selection of the isotope and conjugation procedure. After successful experimental radioimmunoscintigraphy of xenografts (see tables XXVIII, XXIX), the conjugates of two selected antibodies named BW 431/31 and BW 431/26 (both directed against different epitopes of CEA, see table XVI) were tested clinically. Irrespective of whether ^{111}In-labelled F(ab')$_2$ fragments (BW 431/31) or ^{111}In-labelled or ^{99m}Tc-labelled intact immunoglobulins (BW 431/26) were used, the percentage of correct results reproducibly achieved using ECT technology was in the range of 85–90% (see tables XXXII, XXXIII).

In the first study covering 113 patients, MAb BW 431/31 F(ab')$_2$-^{111}In was tested (tables XXXII, XXXIII). A sensitivity of 89% was noted when liver metastases were excluded from examination due to enrichment of ^{111}In in the liver. A sensitivity of 85% was reached after only 24 h and by one single examination. The slight increase to 89% was the result of the combination of several examinations at various times. Compared to the median sensitivity of 77%, calculated in the compilation of 25 studies on the subject of immunoscintigraphy of gastrointestinal tumors [38] (table XXXI), our results seem to be an improvement.

The increasing number of similar investigations shows a growing acceptance of the method of radioimmunoscintigraphy. Iodine itself (^{131}I or ^{123}I) is gradually disappearing into the background. Indium (^{111}In) and especially technetium (^{99m}Tc) are the labelling nuclides gaining importance for routine clinical use.

The wider use of the technique also makes it possible to estimate its diagnostic value. In a recent study by Biersack et al. [54] on an unselected group of 52 patients (mostly with recurrent disease of a variety of malignant tumors) it was proved that in 33% of the cases radioimmunoimaging gave useful additional information. This information was not available through computer tomography, sonography or laboratory tests, and considerably improved the efficacy of treatment in these patients.

All in all, the results of the preclinical experiments in radioimmunolocalization indicate the possibilities as well as the limitations of this technique. Improvement has already been achieved through the development of new methods to stably link radionuclides suitable for radioimmunoscintigraphy such as ^{111}In or ^{99m}Tc without impairing the specificity of the antibody. In addition, there are antibodies already available which are suitable for the immunoscintigraphy of tumors in patients, both with respect to specificity as well as affinity. As a result, we now have at our disposal radiolabelled monoclonal antibodies for clinical routine use in the diagnosis of primaries, local recurrences and metastases of gastrointestinal tumors.

Nevertheless, the antibody preparations should be improved. The aim should be to generate antibodies of a significantly higher specificity, especially for bronchial carcinomas, and of higher affinity; moreover, the smallest possible epitope binding fragment of these antibodies should be prepared to increase the penetration and elimination rate. In addition, stable

Table XXXII. Clinical trial ^{111}In BW 431/31 F (ab')$_2$ — sensitivity of tumor detection (colon carcinoma) in correlation with the time of scintigraphy

Time of scintigraphy	Total number of investigations	(1) Correct positive	(2) Correct negative	(3) Questionable positive	False negative	False positive	Sum of (1)—(3)
24 h p.i.	54 ≙ 100%	33 ≙ 61%	5 ≙ 9%	8 ≙ 15%	8 ≙ 15%	0	46 ≙ 85%
48 h p.i.	60 ≙ 100%	31 ≙ 52%	8 ≙ 13%	11 ≙ 18%	10 ≙ 17%	0	50 ≙ 83%
72 h p.i.	11 ≙ 100%	8 ≙ 73%	0	1 ≙ 9%	2 ≙ 18%	0	9 ≙ 82%
96 h p.i.	24 ≙ 100%	17 ≙ 71%	1 ≙ 4%	5 ≙ 21%	1 ≙ 4%	0	23 ≙ 96%
120 h p.i.	23 ≙ 100%	13 ≙ 57%	3 ≙ 13%	2 ≙ 9%	4 ≙ 17%	1 ≙ 4%	18 ≙ 79%

Applied dose: 1—3 mCi (≙ 1—3 mg MAb) in 50—100 phys. NaCl solution i.v.
* For characteristics of MAb see table XVI

Table XXXIII. a Radioimmunoscintigraphy — clinical results in gastrointestinal tumors

Author	Antigen	Antibody	Isotype	Radio-nuclide	No. of patients	Correct results (%)
Oberhausen et al. (1986)	CEA	BW 431/31	IgG$_1$	^{131}I	15	80
				^{111}In	6	83
Bares et al. (1987)	CEA	BW 431/31	F (ab')$_2$	^{111}In	44	86
Oberhausen et al. (1987)	CEA	BW 431/31 BW 431/26	F (ab')$_2$; IgG$_1$	^{111}In ^{99m}Tc ^{111}In	119	90
Oberhausen et al. (1987)	CEA	BW 431/26	IgG$_1$	^{99m}Tc	22	>90
Baum et al. (1987)	CEA	BW 431/26	IgG$_1$	^{99m}Tc	18	>90

linker systems should be developed which do not impair the epitope binding capacity and can reduce Fc-mediated or any other nonspecific binding of the conjugate to normal tissues. With this strategy, the tumor-to-normal-tissue ratio of radioactivity should be significantly increased and the time necessary to achieve the optimal ratio after application of the antibody should be reduced.

Table XXXIII.b Summary of clinical trials (immunoscintigraphy of colon carcinoma)

	Total no.	(1) Correct positive	(2) Correct negative	(3) Questionable positive	False negative	False positive	Sum of (1)—(3)
Colon carcinoma BW 431/31–^{111}In* (Fab'$_2$)** Liver metastases							
included	113	71 (62.8%)	11 (9.7%)	14 (12.4%)	11 (9.7%)	0	96 (85%)
excluded	108	(66.0%)	(10.0%)	(13.0%)	(6.0%)	0	(89%)
BW 431/26–^{99m}Tc* (including liver metastases)***	22	18	0	3	1	0	21 (> 90%)
BW 431/26–^{111}In***	14	12	0	1	1	0	(90%)

* For characteristics of MAb see table XVI
** Multicenter studies
*** Study by Oberhausen et al., 1987

Specific Radioimmunotherapy

Physicochemical and Biological Considerations

The therapy of malignant tumors with radionuclides conjugated to tumor-specific antibodies appears to be the consequential continuation of the studies which have led to successful tumor immunoscintigraphy.

While in the case of diagnostic immunoscintigraphy the detection outside the body is the only desired effect of radiation caused by the radionuclide, the demands regarding the effect of ionizing radiation from a therapeutic point of view are far more complex. From the diagnostic point of view, radiation doses for the whole body or for parts of the body should be kept as small as possible, while from a therapeutical point of view the highest possible dose in the target area and the lowest possible dose in the rest of the body are the opposing demands made. The quality of radiation must be chosen in such a way that all tumor cells are killed while at the same time the cells of the normal tissue or other structures in the body are not permanently damaged (table XXIV).

These very diverging demands of immunoscintigraphy on the one side and radioimmunotherapy on the other cannot of course be met by one radionuclide alone. For diagnostic use only γ-rays come into question. Ideally these rays should exhibit an energy of 100–200 keV. For tumor therapy, however, radiation must be limited to the tumor only, and for this reason γ-rays are not suitable because of their wide radiation range. Due to their physical qualities, only β- or α-rays come into question. A valuation of radionuclides for radioimmunotherapy is given in table XXXIV. Regarding the range of these rays, we must make sure that the values are not too high and yet strong enough to ensure that the nucleus is reached from the cell membrane. Since the range is energy-dependent, this means that β-rays may at the most produce a maximum energy of 2.5 MeV. In this case the middle radiation range in water would be approximately 4 mm (maximum 12 mm). α-rays, on the other hand, can never reach so far. Therefore, in their case we must demand a minimum energy of 2.5 MeV so that the radiation goes

Table XXXIV. Valuation of radionuclides for radioimmunotherapy of tumors

Parameter	α-emitters	β-emitters
Half-life with respect to time needed by Mab to localize a tumor	too short	suitable
Stability of conjugates	decay to unstable daughter nuclides; rupture of radio-nuclide-antibody ligands	radiolysis of antibody in case of high concentration of immunoconjugate
Organ-specific localization	typical for heavy metals	thyroid gland in case of iodine (can be prevented)
Energy of radiation	high linear energy transfer > 80 MeV$/\mu$m	local energy dependent on range of radiation (the shorter the higher)
Range of radiation	short (50–90 μm)	low ($< 200\ \mu$m) to long (> 1 mm) dependent on radionuclide
Strike probability and effectivity	1–1.5 hits per cell sufficient for cell kill; strike probability $\sim 10\%$	400 hits* are necessary to kill a cell; strike probability $> 100\%$
Repair mechanism	not known (adventageous for the tumor, disadvantageous for normal cells)	available (disadvantageous for the tumor, advantageous for normal cells)
Probability of tumor sterilization in the case of heterogeneous distribution of TAA-positive cells	low (tumor has to be homogeneous with respect to TAA-positive cells; distance should be lower than 50–90 μm)	high (homogeneous radiation of a tumor)

* Humm 1986

further than 15 μm. A list of those radionuclides under discussion for radioimmunotherapy is given in table XXXV.

By rule of thumb, the half-life of a diagnostic radionuclide should be about the length of the examination time. This value is dependent on the kinetics of the antibody and lies somewhere between 4 and 24 h in the case of those antibodies that we have examined up to now (see table XVI). For therapeutic use the kinetics of the MAb play a far greater role. Here it is imperative that the accumulated activity – that is, the integral of activity over time – be as high as possible in the tumor, while being as low as possible in all other structures.

Table XXXV. Nuclides for Radioimmunotherapy

Nuclide	Half-life	Radiation
^{131}I	192.0 h	β; 0.6; 0.8 MeV
^{90}Y	64.0 h	β; 2.3 MeV
^{109}Pd	13.4 h	β; 1.0 MeV
^{67}Cu	62.0 h	β; 0.4; 0.6 MeV
^{186}Re	88.3 h	β; 1.1 MeV
^{188}Re	17.0 h	β; 2.1 MeV
^{211}At	7.2 h	α; 5.9 MeV
^{212}Bi	1.0 h	(64%) β; 2.3 MeV → ^{212}Po (0.3 μs)
		α; 8.8 MeV
		(36%) α; 6.0 MeV → ^{208}TL (3 min)
		β; 1.8; 2.4 MeV
^{255}Fm	20.1 h	α; 7.0 MeV

According to our experience, the maximum tumor accumulation of radiolabelled antibodies is reached after about 24 h. If a large part of the radiation has been emitted up to this time, then this will have led mainly to an undesired body dosage. Only in the case of half-lives which lie clearly over 24 h does the slower decrease in activity in the tumor, in comparison to other tissues, become more effective with respect to and in favor of the accumulated activity in the tumor (see fig. 11). However, in the case of β-emission the half-life must on no account be too long, so that radiation does not become protracted and enable tumor repair processes to take place. Such repair processes are not known of in the case of α-rays. The half-life of therapeutic radionuclides here can therefore lie between 2 and 10 days (see table XXIV).

A further demand is that even after disintegration the nuclide remains stable, because one cannot be sure that the fixation between radionuclide and antibody remains intact after disintegration. Exceptions to this demand are very fast ($<$ 1s) follow-up disintegrations, which end in a stable state. For radiation quality the same demands apply as above.

Where the disintegration of a therapy nuclide results in an accompanying γ-ray, it can be used for immunoscintigraphic therapy control.

Of course, logistical considerations also play a role when choosing radionuclides and a suitable labelling system must be available. However, in this chapter we do not want to probe further into these problems (see p. 42 ff.).

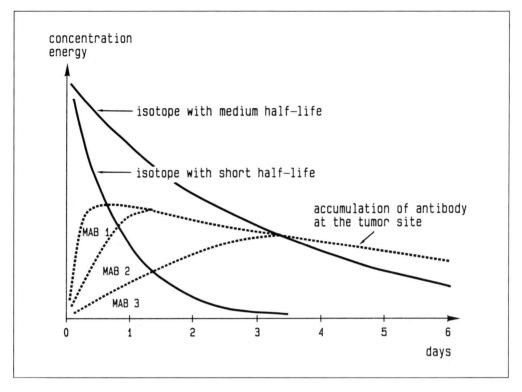

Fig. 11. Relationship between MAb kinetics and half-life of antibody at the tumor site. The graph shows the relationship between the half-life of the isotope and the pharmacokinetics of the monoclonal antibody. The peak of tumor localization of the antibody (MAb 1, 2, 3 are antibodies with different pharmacokinetics of tumor localization) should be reached within a period of time which is shorter than the half-life of the radionuclide conjugated to the antibody. If the necessary conditions are fulfilled, most of the radioactivity linked to the antibody which is localizing the tumor irradiates the tumor.

It is obvious that any preparation suitable for radioimmunotherapy should be tested preclinically not only in vitro but also in vivo. Successful therapeutic effects on xenografted tumors in nude mice do not seem to us to be sufficient for any further development of a radiolabelled antibody preparation for radioimmunotherapy. As has already been discussed (see the previous chapter and table XXVII), the percentage of antibodies localizing to tumors xenografted into nude mice is higher by a factor of about 1,000 than that of antibodies localizing to tumors of human patients. Consequently, results of experimental studies in experimental, mainly xenograft

Table XXXVI. Radioimmunotherapy — clinical trials in cancer patients with monoclonal antibodies

Authors	MAb	Route of application	Disease	Response
Carrasquillo et al. (1984)	8.2 96.5 (p97) 48.7	i.v.	melanoma	2/10 MR
Lashford et al. (1985)	UJ13A	i.v.	neuroblastoma	1/4 MR
Cheung et al. (1986)	anti GD$_2$	i.v.	neuroblastoma	1/3 PR
Epenetos et al. (1987)	HMFG1 HMFG2 AUA1 H17E2	i.p.	ovarian cancer	no response in 7 patients with gross disease 5/6 CR (patients with microscopic disease) 2/13 MR (tumor nodules < 2 cm)
Ashorn et al. (1985)	IIIH2	i.p.	ovarian cancer	1/1 MR
Coakham et al. (1987)	MAbs against neuroectodermal associated antigens	i. cran.	neoplastic meningitis from medulloblastoma, lyphoma, melanoma, pineoblastoma	5/7 PR
Riva et al. (1987)	HMFG1 HMFG2 PLAP AUA1 FO23C5 494/32	i.p./i. pleural	breast cancer colon cancer lung cancer ovarian cancer	1/11 CR 3/11 PR 4/11 SD 3/11 PD
Epenetos et al. (1987)	anti-CEA (431/31)	i. art. hep.	liver metastases of colon cancer	1/1 PR
Mach et al. (1987)	anti-CEA	i. art. hep.	liver metastases of colon cancer	no response in 7 patients
Rosen et al. (1987)	T101		CTCL	6/6 MR

systems, which are favorable to systemic radioimmunotherapy [10, 22, 226, 246, 339, 340, 559, 607, 753] can give us less false-negative but false-positive data. These experiments only confirm the information already known regarding the therapeutic potency of radiation, when the nuclide is concentrated to a sufficient degree at the tumor site. Considering this aspect, positive results in radioimmunotherapeutical experiments with xenografted tumors in nude mice offer only limited or even no predictive value in the clinical situation. What are essential are dosimetric studies on human beings to calculate the risk-versus-benefit ratio of radioimmunotherapy with a selected radiolabelled antibody preparation. These calculations have been performed by us as well as by others [112, 113, 132, 315, 713, 714].

Clinical therapy trials have already been carried out with ^{131}I (table XXXVI). The other nuclides listed in table XXXV are those under discussion for this use.

The iodine isotopes have a special position among the labelling nuclides. The human body has specific enzyme systems at its disposal, the so-called dehalogenase enzymes [550, 754], which cause the iodine to split off the antibody, especially in the organs of the RES (see p. 53 ff.). The freed iodine is then (in the case of a blocked thyroid gland) quickly eliminated renally. This means that the radiation load is reduced in the affected organs.

This example shows how very dependent the distribution of the radionuclide – and that means the radiation dose in the body – is on the kinetics of the antibodies and also on the type of labelling and kind of labelling nuclide.

The same energy dose can produce different biological radiation effects in the tissue, depending on whether these doses are evoked by β- or α-rays. Thus a biological evaluation factor must be taken into consideration (RBE = relative biological effectivity).

The dose evaluation factor used for radiological safety precautions is between 10 and 20 for α-rays; for β-rays, however, it is about 1. Here we must remember that this evaluation is made to fulfill the requirements of radiological safety, in order to be on the safe side of the expected biological effect after exposure to a radionuclide. (It is important not to forget that one single changed cell might be enough for carcinogenesis.) The problem discussed here, however, is on a completely different level: one single undestroyed tumor cell within a living organism means that the tumor continues to exist. For the evaluation of therapeutic success the factor for α-rays thus definitely lies below the values given for radiological safety. For the therapeutic problem under discussion here there is no proof that this qualitative factor cannot lie below the range of 1, i.e. a more unfavorable position than in the case of β-rays.

The probability of striking the nucleus of a cell where the radionuclide emitting α-rays (via the carrier antibody) is fixed to the cell membrane is around 2–4%. Taking the surrounding cells as well as the range of radiation into consideration, then the probability of striking is about 10% for α-rays. However, only 1–1.5 strikes are necessary in order to destroy the cell [5].

For β-rays the probability of striking lies significantly above 100%, since statistically seen every β-particle of the declared energy can reach about 5 nuclei altogether. However, usually more than one strike per unit time is necessary in the target area in order to destroy the cell. In this case the therapeutic effect of the radiation is not only dependent on the accumulated activity but also on the existing activity concentration. The advantage of the β-rays lies in the significantly more homogeneous irradiation of the tumor [315].

A final decision regarding the choice between α-and β-rays cannot be made at present. Rösler and Noelpp [578] classify available α-rays as inferior to the β-rays. The data of Stuart et al. [679], Kurtzman et al. [393] and Black et al. [55] show the effectiveness of the α-emitter ^{212}Bi, conjugated to monoclonal antibodies, in in vitro cytotoxic assays as well as in in vivo assays, but these biological data do not provide sufficient arguments, as discussed above.

The antibody accumulation in the tumor has a key role in the calculation of the dose distribution for a therapy trial. The value reached in humans (and in many cases not even achieved) is on average 0.01%/g tumor in the case of systemic application of the MAb. This means that 0.01% of the dose of the i.v. injected antibodies can be found in 1 g of tumor (table XXXVII) [135, 184, 400, 426].

Using this value, we made dosimetric calculations in patients. We can, for example, calculate for anti-CEA MAb BW 431/31 the dose distributions after intravenous injection of 4 GBq ^{131}I–MAb (see tables XXXVII, XXXVIII).

The dose of this applied activity is about the same size as is commonly used for the therapy of thyroid gland carcinoma in patients. Due to the deiodinization process in the RES the highest dose values lie in the tumor. Nevertheless, only 8 Gy/g tumor can be reached. This dose is far below the local dose of 80 Gy at which a therapeutic effect on the tumor can be expected [315]. In contrast to this, the dose which accumulates in the bone marrow is near the maximal tolerated dose of about 2 Gy.

Table XXXVII. Systemic radioimmunotherapy with IgG or F(ab')$_2$ fragments — dose calculation after i.v. application of 4 GBq ^{131}I MAb (108 mCi)*

Whole body (inclusive bone marrow)	1.3 Gy**
Liver	2.7 Gy
Spleen	2.3 Gy
Kidney	2.2 Gy
Tumor	8.0 Gy***

* Calculation is based on the assumption that 0.01 % of total antibody dose is localized/g tumor
4—8 GBq is the commonly used activity in the treatment of relapses and metastases of thyroid gland carcinoma
** Low tissue values by enzymatic cleavage and excretion of ^{131}I
*** Therapeutic effect is in the range of about 80 Gy
1 Gy = 100 rad; maximal tolerated dose for bone marrow ≙ 2 Gy

Table XXXVIII. Systemic radioimmunotherapy with IgG or F(ab')$_2$ fragments — dose calculations after i.v. application of 2 GBq ^{90}Y MAb* (calculated according to dosimetry with ^{111}In MAb)

Liver	6.0 Gy
Spleen	38.0 Gy
Kidney	52.0 Gy
Tumor	8.0 Gy**

* Calculation is based on the assumption that 0.01 % of total antibody dose is localized/g tumor
** Therapeutic effect is in the range of about 80 Gy
1 Gy = 100 rad

Even more extreme are the relations in the case of stable labelling where no effect takes place which corresponds to deiodinization. Table XXXVIII shows dose values which were obtained by transferring kinetic data obtained with ^{111}In-BW 431/31 onto yttrium (^{90}Y-BW 431/31). In order to reach the same tumor dose of 8 Gy/g (as for example ^{131}I-BW 431/31) one must inject 2 GBq ^{90}Y–MAb. The resulting kidney value lies more than a factor of 6 above the tumor dose and is in the range of the maximal tolerated dose.

The critical organ is the bone marrow. But just here it is very difficult to quantify examinations on humans. The dose applied to the bone marrow after systemic therapy with ^{90}Y–MAb is certainly higher than the one for the liver and probably nearer to that for the spleen. In each case, it is above the maximal tolerated dose for the bone marrow. Consequently, in the course of systemic radioimmunotherapy one has to expect considerable irreversible damage to the bone marrow. The suggestion made by Larson et al. [400] to take out bone marrow before radioimmunotherapy and to replant it after therapy does not appear feasible to us in routine clinical work.

Both of the above-mentioned examples show that, in the case of tumor accumulation of the kind that can be achieved today using antibodies in patients, it does not seem possible to us to carry out radioimmunotherapy of tumors through the systemic application of radiolabelled antibodies. This corresponds to the judgement of other authors [132, 184, 315, 468, 713, 714] and supports the application of radiolabelled intact IgG or F(ab')$_2$. By contrast, Carrasquillo et al. [112] claimed completely different dosimetric data regarding the application of Fab fragments: for every 100 mCi of ^{131}I Fab administered they calculated 10 Gy for the tumor, 0.3 Gy for the liver and 0.03 Gy for the bone marrow. They claimed that up to 342 mCi could be repetitively injected i.v. into melanoma patients without excessive organ toxicity. Further studies are thus needed to evaluate whether the kinetics of Fab fragments, and especially those of ^{131}I–Fab fragments, are completely different from those of intact IgG or F(ab)$_2$ fragments.

In contrast to systemic application, the regional (intracavitary or intratumoral) application of IgG or F(ab')$_2$ fragments may result in higher tumor retention values [478] of the radiolabelled antibodies, so that a significantly more favorable dose distribution can be reached.

Clinical Trials

Some clinical trials have already been performed to evaluate the effect in tumor therapy of systemically or regionally applied radiolabelled antibodies. In spite of the unfavorable dosimetric evaluation, some of the early clinical phase-I studies of systemic radiolabelled antibody therapy of tumors have produced positive results. It is, however, inherent in a phase-I study that positive results in tumor therapy have to be looked upon with caution.

The most extensively examined antibody for radioimmunotherapy was ^{131}I-labelled antiferritin heteroantisera, in a series of studies by Order et al.

[514] and others [190, 407, 408]. These studies utilized antisera prepared in a variety of species and which were administered to the patients in a cyclic manner to avoid the development of an immune response to the infused antibody. Radiation doses to the tumor were in the range of 1,000–1,500 cGy after infusion of 50 mCi in 2 divided doses. About 50% of the patients treated achieved a partial response documented by tumor volume measurements. Toxicity was primarily hematologic due to β-emission of ^{131}I. However, since the use of ^{131}I-labelled antibody was part of a combined treatment regimen including chemotherapy, the exact contribution of radiation to the observed clinical improvement is difficult to determine.

Subsequently, Lenhard et al. [410] have extended these studies of ^{131}I-labelled antiferritin to the treatment of Hodgkin's disease, a tumor that also contains ferritin. The results in these patients were similar to those with hepatoma. After i.v. infusion of 50 mCi of ^{131}I-labelled antiferritin antibody, 40% of patients had an objective response with 1/37 patients achieving a complete response.

Smaller numbers of patients were treated with ^{131}I-labelled MAbs. Larson et al. [397, 398] and Carrasquillo et al. [112] have used Fab fragments of a MAb directed against the melanoma-associated antigen p97 for radioimmunotherapy of metastatic melanoma. Here large doses of radiolabelled antibody (up to 342 mCi) could be administered repetitively (i.v.) to patients without excessive organ toxicity. Two of 10 patients treated with high-dose radiolabelled antimelanoma Fab showed an effect from the treatment – however, no complete or partial response. Irrespective of dosimetric considerations, it is most likely that the relatively high radioresistance of melanoma, as compared to hepatoma or Hodgkin's disease, may contribute to the lack of effectiveness.

Recently, Lashford et al. [401] used ^{131}I-labelled MAb UJ 13A in 4 children with disseminated neuroblastoma. After i.v. infusion of 35–55 mCi 1 patient had an objective response. Of particular note in this study is 1 patient who received 55 mCi and who developed irreversible marrow aplasia although the estimated dose of radiation to the marrow was 66 cGy. Similar responses could be observed [127, 128] with an anti-GD2 MAb labelled with ^{131}I.

In comparison with these slight and/or uncertain effects of, and side effects induced by, systemic radioimmunotherapy of tumors, the regional application of radiolabelled monoclonal antibodies resulted in probably more significant effects in tumor therapy and lower toxicity (see table XXXVI).

At the Hammersmith Hospital in London 34 patients with resistant ovarian cancer were treated by intraperitoneal – which means local – administration of ^{131}I-labelled antibodies (HMFG1, HMFG2, AUA1, H17E2 [185]). It was shown that tumor uptake was 5 times greater than after i.v. administration. There were no significant responses in 7 patients with gross disease. However, out of 6 patients with a microscopic disease, 5 are disease-free with target follow-ups being carried out over 3 years. There were also 2 responses in 13 assessable patients with tumor nodules of 2 cm in diameter.

A new phase I–II study has been initiated at the same hospital using ^{90}Y as the radiolabel.

Similar therapeutic approaches in ovarian cancer using anti-HMFG MAbs have been tested by others [18].

That regional application of radiolabelled antibodies may induce better therapeutic responses has also been indicated by other studies. Coakham et al. [131] demonstrated, for example, that a subarachnoid injection of a combination of ^{131}I-labelled MAbs (11–50 mCi) resulted in biological responses and increased survivals (1–2 years) in 5 of 7 patients with neoplastic meningitis from lymphoma, melanoma, medulloblastoma and pineblastoma resistant to conventional therapy.

In addition, Riva et al. [575] reported on 11 patients suffering from advanced cancer, whereby in 8 cases MAbs could be injected via an intraperitoneal or an intrapleural route (50–150 mCi) instead of an i.v. route. The calculated dose delivered to the tumor was between 1,800 and 5,200 rad. One patient responded with a complete remission, 3 with partial remission, 4 with static disease and 3 with progressive disease. In the case of 1 patient with liver metastases of colorectal cancer, Epenetos et al. [185] demonstrated that combined injection of anti-CEA MAbs and microspheres into the liver arteria induced a long-term partial remission.

However, a similar approach of regional therapy used by another group failed in 7 patients [431].

Good candidates for radioimmunotherapy with MAbs may be leukemias and lymphomas because of their radiosensitivity. In a study in patients with cutaneous T-cell lymphoma all 6 patients responded to an initial therapy with ^{131}I MAbs T101 and 2 patients responded to retreatment [581]. All patients reported resolution of their chronic pruritus. The duration of response ranged from 3 weeks to 3 months.

The results of phase–I studies seem to suggest that radiolabelled antibodies might be potentially useful in treating a malignant disease, at least in the case of regional application.

However, this conclusion is too early. Similar to the situation with other new compounds for tumor therapy [603], any kind of conclusion concerning the therapeutic effect of regional radioimmunotherapy can only be made in case the tumor regressions observed in phase-I studies can be confirmed by subsequent and adequately controlled randomized studies.

With respect to systemic radioimmunotherapy, the results of our dosimetric studies as well as of others strongly argue against the possibility that the regressions of solid tumors being recorded in the above-mentioned phase-I studies after intravenous application of radiolabelled antibodies may be caused by radiation. If Fab fragments do not behave differently from IgG or F(ab)$_2$ and if no drastic improvement in the amount of the antibody localizing at the tumor site can be made and/or if the amount of antibody or radionuclides localizing in normal tissues cannot be reduced significantly by new technology, we believe that the risk-versus-benefit ratio resulting from our dosimetric study does not justify any further clinical study for systemic radioimmunotherapy in solid tumor diseases.

Summing up, we can conclude that the low localization rate at the tumor site of IgG and F(ab')$_2$ antibodies is the main obstacle against using radiolabelled monoclonal antibodies in systemic tumor immunotherapy. Future research must concentrate on improving this localization rate to increase the effect of therapy on the tumor and to reduce side effects (see table XXXIX). The chances of success might be increased by generating and selecting antibodies specific for epitopes on TAA with a high tumor selectivity and a high density per cell, or by increasing the avidity of the antibody to a level higher than 10^8 l/mol, or by the preparation of antibody fragments even smaller than Fab fragments which hopefully are able to penetrate quickly extravascularly to the tumor site and which in addition are excreted by the kidney at high speed, or by the development of other procedures for rapid elimination of non-tumor-bound antibodies and/or for reduction of the localization of the radiolabel to normal tissues.

The selection of the radionuclide should depend on the results of the further development of conjugation methods, which should be chosen according to their effect on the in vivo behavior of the immunoconjugate.

Obviously, this is a risky research project. Alternatively or additionally the further development of techniques and devices for applying radiolabelled monoclonal antibody regionally to certain tumor-diseased organs may also be very helpful in evaluating the potency of this type of treatment in tumor diseases. The results of clinical studies performed so far, in which radiolabelled antibodies have been applied locally, speak in favor of this development.

Table XXXIX. Research lines for improving parameters to enable systemic radioimmunotherapy

---

Selection of TAA
- Sufficient amount on most of the tumor cells
- Homogeneous distribution within the tumor
- Amount $> 10^6$/cell

Selection of antibody
- High-avidity antibodies
  $K_a \geq 10^9$ l/mol
- Fragment with extremely low unspecific binding, high penetration capacity and short half-life

Conjugation method for β-emitters (or α-emitters)
- Selection of nuclides and development of conjugation methods which enable quick elimination of antibodies not bound to the tumor and/or with a drastic reduction of localization of antibodies in normal tissue

---

Specific Chemoimmunotherapy

Initial studies stemming from the work of Mathé et al. [445] upon the synthesis of drug (aminopterin) antibody conjugates relied exclusively on the use of conventional polyclonal antibodies. The experience gained so far has provided a basis for the development of conjugates using monoclonal antibodies, mostly of murine origin [221–223]. The aim has been to conjugate a maximum number of drug residues to antibodies under conditions ensuring optimal retention of both drug and antibody moieties [57, 223]. It is obvious that preservation of the activity and affinity of the carrier antibody is essential for its targeting function. This is difficult to achieve, because slight denaturation of the antibody during labelling may shorten the half-life of the antibody [541]. Cytotoxic agents conjugated to the antibody may alter the conformation of the combining site, especially if the agent is distinctly hydrophobic, or possess multiple charged groups, or they may produce steric hindrance of antigen binding [238].

These effects might correlate directly with the molar ratio of cytotoxic agents bound to the antibody [235–237, 541]. F(ab)$_2$ fragments seem to be more sensitive for inactivation than intact IgG [388].

Conjugation of Cytostatics or Toxins

In general, reactive groups in Ig potentially available for linkage of cytostatics or toxins exist in the side chains of the amino acids and in the carbohydrate moieties. They include aliphatic carboxyl, amino, disulfide and hydroxyl groups; imidazole and phenolic rings and aromatic hydroxyl groups (table XL). The most widely used thus far have been carboxyl and amino groups [238].

The various methods of linking cytotoxic agents to immunoglobulins directly or via heterobifunctional coupling have been reviewed recently [57]. The main procedures for the linkage of compounds to the amino, sulfhydryl and carboxy group of proteins are listed in figure 12.

Noncovalent Binding

Target (TAA) ⇌ [IgG – Biotin ⇌ Avidin – S – S – Toxin (A chain)]	Hashimoto et al. (1984)
Target (Fc-Rec.) ⇌ [IgG ⇌ Toxin (A chain) Immune Complex]	Shen et al. (1984)
Target (TAA) ⇌ [IgG (Hybrid Antibody)] ⇌ Toxin (A chain)	Raso et al. (1982)

Covalent Binding

A) Linkage to Protein-NH_2

Structure	Reagents	Reference
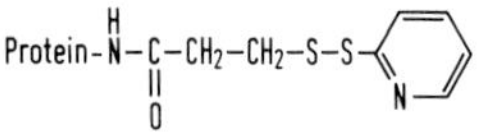 Protein-NH-CO-CH_2-CH_2-S-S-(2-pyridyl)	N-succinimidyl-3-(2-pyridylthio)-propionate (SPDP)	Carlson et al (1978)
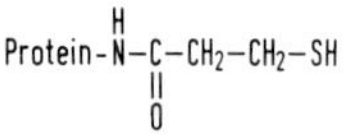 Protein-NH-CO-CH_2-CH_2-SH	SPDP + dithiothreitol	Carlson et al. (1978)
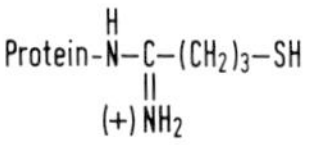 Protein-NH-C(=NH_2^+)-$(CH_2)_3$-SH	2-iminothiolane	King et al. (1978)
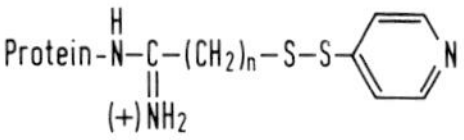 Protein-NH-C(=NH_2^+)-$(CH_2)_n$-S-S-(pyridyl)	n = 2 2-iminothiolane + 4,4 dithiopyridine n = 3 methyl 3-(4'-dithiopyridyl) mercaptopropionimidate	King et al. (1978)
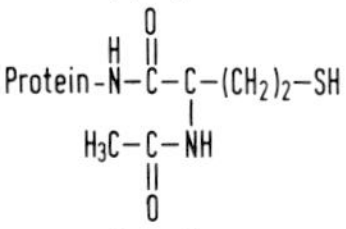 Protein-NH-CO-CO-$(CH_2)_2$-SH (H_3C-CO-NH-)	N-acetylhomocysteine thiolactone	Reiner et al. (1977)
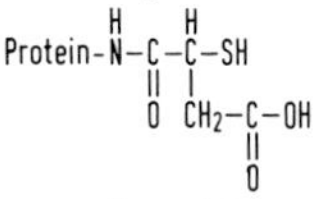 Protein-NH-CO-CH(-CH_2-COOH)-SH	acetylmercaptosuccinic anhydride + NH_2OH	Klotz and Heineg (1962)
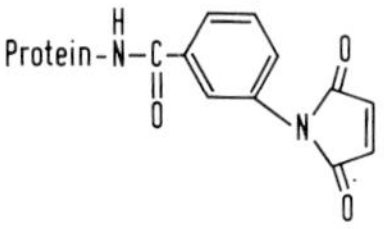 Protein-NH-CO-(phenyl-maleimido)	m-maleimido-benzoyl-N-hydroxysuccinimide ester	Lin et al. (1979)
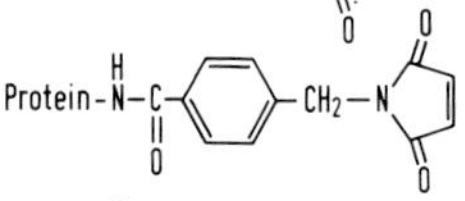 Protein-NH-CO-(phenyl-CH_2-maleimido)	succinimidyl 4-(N-maleimido-methylcyclohexane)-1-carboxylate	Yoshitake et al. (1979)
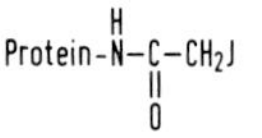 Protein-NH-CO-CH_2J	N-succinimidyliodoacetate	Rector et al. (1978)
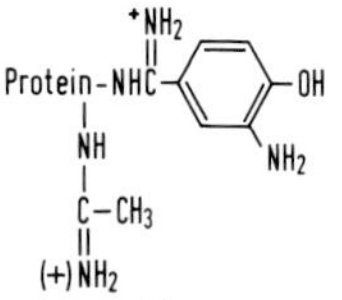 Protein-NHC(=NH_2^+)-(phenyl-OH, -NH_2) (NH-C(=NH_2^+)-CH_3)	4 hydroxy 3-nitromethyl-benzimidate + acetimidate + $Na_2S_2O_4$	Müller and Pfleiderer (1979)
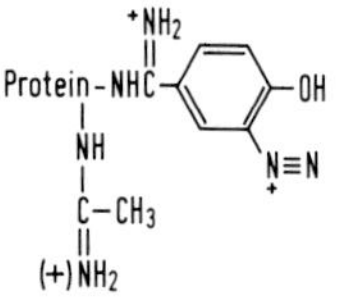 Protein-NHC(=NH_2^+)-(phenyl-OH, -N≡N$^+$) (NH-C(=NH_2^+)-CH_3)	4 hydroxy 3-nitromethyl-benzimidate + acetimidate + $NaNO_2$	Müller and Pfleiderer (1979)
Protein-NH_2	oxidized dextran + borohydride	Hurwitz et al. (1978)

B) Linkage to Protein-Hydroxy Group		
Protein-OH	cystamine carbodiimide	Erlanger et al. (1967) Gilliland (1980)
C) Linkage to Protein-SH		
Protein$_1$–S–S–Protein$_2$	Protein-SH	Ghose and Blair (1987)
Protein$_1$–S–SO$_3$Na	Protein-SH + Na$_2$S$_4$O$_6$	Masuko et al (1979)
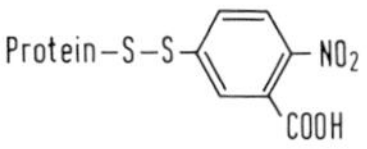 Protein–S–S–C$_6$H$_3$(NO$_2$)(COOH)	Ellman's reagent	Raso and Griffin (1980)
D) Linkage to Protein-Aldehyde Groups		
Protein–CHO	periodate	Hurwitz et al. (1975)
E) Linkage to Protein-COOH		
Protein–C(=O)–NH–CH$_2$–CH$_2$–S–S–CH$_2$–CH$_2$–NH$_2$	cystamine + carbodiimide	Gilliland (1980)

Fig. 12. Conjugation methods for proteins (pp. 82, 83).

To prevent deterioration or hindrance of antigen binding by random linkage of cytotoxic agents to the idiotype of the antibody, conjugation procedures have been developed which use those linkage groups in the IgG which are likely to be absent from the antigen binding site [57, 235–237]. As is the case in the conjugation of radionuclides (see p. 42ff.), the carbohydrate prosthetic group is usually present in the constant region of heavy chains but only occasionally in the variable region. The amount of carbohydrate varies from 3% (in IgG) to 13% (in IgE) and contains galactose, fucose, sialic acid and mannose. Because of their hydrophilic properties and bulk they may be accessible to linkage reagents. The other linkage side can be free sulfhydryl groups formed on reductive cleavage of interchain disulfide bridges in immunoglobulins – e.g. by treatment with low-molecular-weight thiols (mercaptoethanol and dithiothreitol).

The conjugation scheme may require initial modification of either the immunoglobulin or the cytotoxic agent, including the introduction of new functional groups, the incorporation of spacers and intermediates [57].

Table XL. Direction of spacer-mediated binding of cytostatics to antibodies

Binding to carboxyl groups	
Methotrexate	Chu and Whiteley, 1977; Kulkarni et al., 1981; Mathé et al., 1958; Gallego et al., 1984; Kulkarni et al., 1985; Uadia et al., 1985
Chlorambucil	Ross, 1975; Tai et al., 1979
Adriamycin	Hurwitz et al., 1975
Daunomycin	Hurwitz et al., 1980
Neocarcinostatin	Kimura et al., 1980
Binding to amino groups	
Adriamycin	Hurwitz et al., 1975; 1978
Daunomycin	Lee et al., 1978
Methotrexate	Belles-Isles and Pagé, 1981; Tsukada et al., 1982; Gallego et al., 1984; Arnon et al., 1982; Pimm et al., 1982; Deguchi et al., 1987; Garnett et al., 1983
Binding to hydroxyl groups	
Cytosine arabinoside	Erlanger et al., 1967; Moss et al., 1963
5-Fluorouracil	
Adriamycin	Hurwitz et al., 1975
Daunomycin	
Melphalan	Smyth et al., 1987
Methotrexate	Kanellos et al., 1985
Vindesine	Rowland et al., 1985; 1986; Pimm et al., 1984; Embleton et al., 1983
Binding to carbohydrate groups	
Daunomycin	Hurwitz et al., 1980
Binding to sulfhydryl groups	
Trenimon	Linford et al., 1974
Diphtheria toxin	Moolten and Cooperband, 1970; Philpott et al., 1973; Thorpe et al., 1981
Ricin	Youle and Neville, 1980;
Ricin A chain	Carlson et al., 1978
Diphtheria A chain	Masuho et al., 1979
Non-covalently binding	
Chlorambucil	Ghose and Nigam, 1972; Ghose et al., 1983; Ross, 1975

Homo- or heterobifunctional groups can produce cross-links between carboxy, hydroxy, sulfhydryl and/or amino groups of cytotoxic agents and the immunoglobulin with or without bridging by intermediates. Intermediates with series of potentially reactive groups have been selected to increase the number of cytotoxic molecules without impairing their function (table XLI). Thus 50–100 cytostatic molecules can be linked to one molecule of dextran [317, 586] or albumin [224] or polyglutamic acid [585], and one of these complexes is subsequently bound to one immunoglobulin molecule.

Analysis and Parameters of Antitumoral Activity

The number of cytostatic agents as well as the positions of the functional groups on the surface of the immunoglobulin molecule to which the cytostatic agents are either directly or indirectly linked are randomly selected. Consequently, the final immunoconjugate will be endowed with a certain degree of inherent heterogeneity with respect to conjugation ratio, molecular size charge and other physicochemical parameters as well as to

Table XLI. Linkage of cytostatics via multivalent intermediaries

Cytostatic	Intermediary	Authors
Adriamycin	dextran	Ghose et al., 1983; Hurwitz et al., 1978
Daunomycin	dextran	Pimm et al., 1982; Deguchi et al., 1987; Arnon and Sela, 1982
Bleomycin	dextran	Manabe et al., 1983
Mitomycin C	dextran	Manabe et al., 1985
Methotrexate	dextran	Manabe et al., 1984
Arabinocytosine	dextran	Hurwitz et al., 1985
5-Fluorouracil	dextran	Hurwitz et al., 1985
Chlorambucil	dextran	Rowland, 1977
Methotrexate	albumin	Garnett et al., 1983
Para-phenylendiamine mustard	polyglutamic acid	Rowland et al., 1975; Tsuhada et al., 1984
Daunomycin	polyglutamic acid	Kato et al., 1984
Arabinocytosine	polyglutamic acid	Kato et al., 1984

cytostatic activity and antibody function. This implies difficulties in the purification and biochemical standardization of immunoconjugates [238]. On the other hand, the biological activity of the immunoconjugate can only be averaged.

Linkage of cytostatic agents to immunoglobulin may either augment or inhibit the cytostatic activity. For instance, linkage of chlorambucil preserves the alkylating activity and the cytotoxic action [258]. Loss of activity has been reported after conjugation of methotrexate [346, 386–388], whereas Garnett et al. [224] described similar or even increased cytostatic activity of the immunoconjugate, compared to free methotrexate.

In experiments with trenimon [234, 416], purothionin [324], adriamycin [232, 235–237], and vindesine [179] it was observed that the free drug was superior to the immunoconjugate with regard to its cytostatic activity. By contrast, in other trials with daunorubicin [16, 17, 44–46] and vindesine [338] the immunoconjugate was superior to the free drug.

Irrespective of whether conjugation of the cytostatic drug to an antibody might impair its cytotoxic activity, two essential conditions have to be fulfilled by immunoconjugates in order to be used as a carrier-specific cytotoxic drug: Firstly, the antigen-specific binding of the antibody should be preserved; secondly and more importantly, the cytotoxic activity of the immunoconjugate for tumor cells exposing the corresponding specific TAA should be significantly higher than the cytostatic activity of the respective cytostatic drug alone.

Antigen-specific binding can be measured semiquantitatively, either through cytological immunofluorescence in the double antibody assay by titration of the immunoconjugate (being the first antibody in this test system) or, better still, in a direct assay, in which the percentage of a constant amount of ^{131}I-labelled immunoconjugate bound to an excess of cells expressing TAA is quantitatively evaluated in the same manner as has already been described for radiolabelled antibodies (see pp. 49ff.).

The cytostatic activity should preferentially be measured in vitro first; the cloning assay [382] can be used. It is independent of transient toxic effects on the cell metabolism and enables us to identify the drug concentration which destroys 50% of the tumor cells after either short-time (1 h) or long-time (> 24 h) exposure. Other conventional methods, however, have also been used (for review see [238]). In the case of cytostatics, the prerequisite of maintained antibody specificity after the conjugation of chlorambucil, methotrexate, anthracyclines, vindesine, cytosin-arabinoside, 5-fluorouracil, bleomycin or neocarcinostatin was fulfilled by a number of investigators [179, 221–223, 362, 363, 436–438, 681]. The prerequisite of a significant increase in specific cytotoxicity was fulfilled as well (for instance, by [45, 50, 221, 222, 320, 352–354, 386, 387, 437, 438, 587]). The application of such characterized and effective immunoconjugates (between antibodies and chlorambucil, methotrexate, daunomycin, adriamycin, vindesine, melphalan, mitomycin C or neocarcinostatin) either i.p. or i.v. in mice which had been s.c. or i.p. transplanted with a tumor exhibiting the TAA corresponding to the antibody specificity of the immunoconjugate resulted, in nearly all experiments, in a specific therapeutic effect on the tumor which was significantly superior to the effect noted after application of the respective cytostatic compound alone [16, 17, 45, 149, 233, 234, 346, 352, 353, 362, 363, 437, 438, 529, 624, 625, 703]. In these cases, where the immunoconjugate exhibited in vitro cytotoxic activity which was not stronger than the respective cytostatic compound, similar antitumoral activity of both compounds could also be found in vivo [530]. These data together indicate a direct correlation between in vitro and in vivo activity of immunoconjugates (see table XLII). Exceptions to this rule, however, were found in some experiments where antibodies had been conjugated with methotrexate either directly [231, 388] or via the intermediate HSA [180, 224], or where they had been conjugated with daunorubicin via dextran [319, 320].

In all these experiments, the in vitro inhibition of dehydrofolate reductase (DHFR) or the target-specific in vitro cytotoxicity of the immunoconjugate was less than that of the free drug. In spite of this, the in vivo antitumoral activity of the immunoconjugate revealed itself as superior to the drug alone, partly due to a reduced toxicity and a higher maximal tolerated dose of the immunoconjugate [180, 388].

Immunoconjugates have to be incorporated into the target cell to act cytotoxically. When taking into consideration the mechanisms of the action of alkylating agents, methotrexate and vindesine, this assumption seems rational. In the case of adriamycin (ADM) Ghose et al. [236, 237] showed that after conjugation to immunoglobulin or dextran the bindings to and intercalation into DNA of ADM is lost. After uptake of the conjugate by the target cell, however, anthracyclines are cleaved from the immunoconjugate by lysosomal enzymes and can thus act cytotoxically on the DNA [466, 613, 614, 700]. In addition, however, a direct cytotoxic activity of immunoconjugates interacting with the cell membrane of the target cell has to be considered [661, 699].

After linkage of the cytostatic drug to the immunoglobulin, the target-specific toxicity of the resulting immunoconjugate depends on its relative

Table XLII. Experimental chemoimmunotherapy with immunocytostatics

Antibody	Cytostatic	Drug/antibody ratio	AB binding (%)	Exposure time (h)/test system	Cytotoxicity in vitro activity	specificity	Antitumoral activity in vivo (application site, tumor/drug)	Author
Anti-FP	DMN	3–4	100	24/T	≦ DMN	+	> DMN (i.p./i.p.)	Tsukada et al., 1982
Anti-moloney	DMN	25–30	75	3/U	≦ DMN	+	> DMN (i.p./i.v.)	Arnon et al., 1982
Anti-CEA (polyclonal)	DMN	2.0	?	1/T	>> DMN	+	> DMN	Belles-Isles and Pase, 1980
Anti-SP 4 (breast carcinoma)	ADM	19–27	100	24/T 0.5/i.v.	>> ADM	+	> ADM (s.c./i.p.)	Pimm et al., 1982
Anti-prost. AP	ADM	27–100	90	20/U	< ADM	?	≳ ADM (s.c./i.v.)	Deguchi et al., 1987
Anti-Ly-2.1	Melph.	10–20	100	24/T	>> Melph.	+	> Melph. (s.c./i.v.)	Smyth et al., 1987
Anti-Pass. Mel.	Chloramb. (CA)	~1.5	100	0.5/Tp	> AB+C'	+	> AB; CA (i.p./i.p.)	De Weger et al., 1982
Anti-SL-2Ly	Chloramb. (CA)	~1.5	100		> AB+C'	+	> AB; CA	De Weger et al., 1982
Anti-H6-Hep.	Treminon (T)	1–4.5	100	24/Tp	< T	(+)	> T (s.c./i.v.)	Ghose et al., 1982
250–30.6 (colon cancer)	MTX	8–10	100	0.5/U 24/U	< MTX	+	> MTX; AB (s.c./i.p.)	Kanellos et al., 1985
A3CG (TFR)	MTX	8–10	100		< MTX	+	> MTX; AB	Kanellos et al., 1985
Anti-Ely-Ly	MTX	5–12	70	DHFR	< MTX	(+)	≦ MTX (i.p./i.p.)	Kulkarni et al., 1985
96.5 (Mel.)	VDS	3–9	100	0.25/C 1/C 6/C	≧ VDS	+	≧ VDS (s.c./i.p.)	Rowland et al., 1985; 1986

791 T/36	VDS	3–9	100	≧ VDS	+	≧ VDS
11285 (CEA)	VDS	3–9	100	≧ VDS	+	≧ VDS
149555 (CEA)	VDS	3–9	100	≧ VDS	+	≧ VDS
11285 (CEA)	VDS	3–9	100	≧ VDS	+	≧ VDS
791 T/36	VDS	3–6	100	= VDS	+	= VDS (sc/i.v.)

Pimm et al., 1984

T	= ^{3}H-thymidin incorporation	Melph.	= melphalan	
U	= ^{3}H-uridine incorporation	Chloramb.	= chlorambucil = CA	
i.v.	= in vivo tumorigenic test	MTX	= methotrexate	
Tp	= trypan blue exclusion test	VDS	= vindesine	
C	= cloning assay	DHFR	= dehydrofolate-reductase	
DMN	= daunomycin	AB	= antibody	
ADM	= adriamycin	C'	= complement	

FP	= fetoprotein
SP	= Schwangerschaftsprotein
AP	= alkaline phosphatase
TFR	= transferrin receptor

concentration, on the number of the target molecule and on the equilibrium constant governing the interaction between them. Thus in vitro as well as in vivo the dose-dependent cytotoxicity of the immunoconjugate on the target cells should be influenced directly by the cytotoxic potency of the drug, the affinity of the antibody to which the drug is linked and the number of TAA exposed by the tumor cell. Improvements in the specific cytotoxic effect of the immunoconjugate may thus be achieved by the selection of better antibodies, by optimal conjugation methods which do not impair the function of the antibody or that of the cytostatic drug and, last but not least, by the selection of drugs possessing an extremely strong cytotoxic potency. The cytotoxic drug with the highest molar activity seems to be adriamycin. According to our intracellular measurements (Hoffmann, Kraemer, Sedlacek, unpublished results), about 2×10^7 molecules of adriamycin are needed in order to kill an adriamycin-sensitive leukemia cell (L1210) with a 50% certainty. Ten times more molecules of adriamycin are calculated to be necessary to achieve a 99% certainty of cell destruction. In view of this limitation, other drugs more highly active cytostatically than adriamycin were selected for conjugation to immunoglobulins. In consequence, bacteria or plant derived toxins with a lethal activity of about 1–100 molecules/cell were selected (see table XLIII) for linkage to antibodies [114, 115, 177, 742]. In the main, those polypeptides were selected which consisted of an A chain being the cytotoxic entity, and a B chain being a lectin responsible for binding to cell surface carbohydrate moieties. The attachment of the lectin to the cell membrane causes a pinocytosis of the whole A chain-B chain polypeptide complex by the target cell and thus a transfer of the A chain into the cytoplasm. The transfer into the cell is essential for the cytotoxicity of the A chain. Once located intracellularly, the A chain can display its toxic effect on the host cell (see fig. 13). In the case of ricin A chain or of diphtheria toxin A chain, this toxic effect is mediated by catalyzing an inactivating reaction on ribosomes and on the elongation factor. To get rid of the general toxicity of the bacteria and plant polypeptide toxins (in the case of ricin toxin, every cell exposing galactose-containing receptors for the B chain lectin is sensitive to ricin cytotoxicity) the B chain was cleaved from the A chain and substituted by a specific antibody. Compared to intact toxin the isolated ricin A chain is less cytotoxic in the order of 5–6 logs [479, 536], but it recovers its cytotoxicity by linkage with an IgG antibody [114, 115, 332]. A prerequisite for the toxicity of this immunoconjugate, however, is its endocytosis by the respective target cells. This endocytosis may be mediated by specific binding of the immunoconjugate to tumor-membrane-associated TAA.

Table XLIII. Examples of toxins, linked to monoclonal antibodies

Toxin	Target	Author
Diphtheria toxin A chain	SW1116 colorectal cancer cells	Gilliland et al., 1980
Ricin A chain	SK-MEL28 melanoma	Casellas et al., 1982
PAP	B-ALL cells	Uckun et al., 1985
Gelonin	AKR-A mouse lymphoma cells	Thorpe et al., 1981
Pseudomonas exotoxin	A431 lung carcinoma cells	Fitzgerald et al., 1983
Saponarin	AKR-A mouse leukemia cells	Thorpe et al., 1985
Diphtheria toxin	Daudi lymphoma cells	Thorpe et al., 1978
Abrin	AKR-A mouse leukemia cells	Thorpe and Ross, 1982
Ricin	EL4 mouse leukemia cells	Youle and Neville, 1980
Abrin A	L10 guinea pig hepatocarcinoma cells	Hwang et al., 1984
Pokeweed antiviral protein	B-acute lymphatic leukemia cells	Uckun et al., 1985 a,b

The target-specific intracellular cytotoxicity of ricin A chain immunoconjugates may be intensified by the simultaneous application of ricin B chain immunoconjugates of identical antibody specificity [715–717]. The ricin B chain is known to augment internalization of conjugates [497, 749, 750].

Binding to TAA expressed by any other normal cell or binding of the immunoconjugate via its Fc part to any normal cell exposing Fc receptors may result in pinocytosis and destruction of that normal cell, causing significant side effects. The Fc receptor binding of the immunoconjugate may be avoided by the conjugation of A chain toxin fragments to Fab or F(ab')$_2$ fragments instead of intact IgG [716]. These Fab or F(ab')$_2$ immunoconjugates exhibit the same cytotoxic activity to the corresponding target cells as the intact IgG conjugates.

As in the case of immunocytostatics, most, if not all of the immunotoxins which were therapeutically effective in experimental tumor systems fulfilled the following conditions: the antibody specificity and binding activity was not impaired by the conjugation method, and in vitro the immunoconjugate significantly exhibited higher target-specific cytotoxic activity compared to ricin A chain (see table XLIV) [58, 200, 701].

Similar to the case of radioimmunotherapy, it is questionable whether the results of therapeutical experiments with immunocytostatics or immunotoxins in human tumor xenograft models have a predictive value in patients. It is not only a matter of controversy whether xenografted tumor

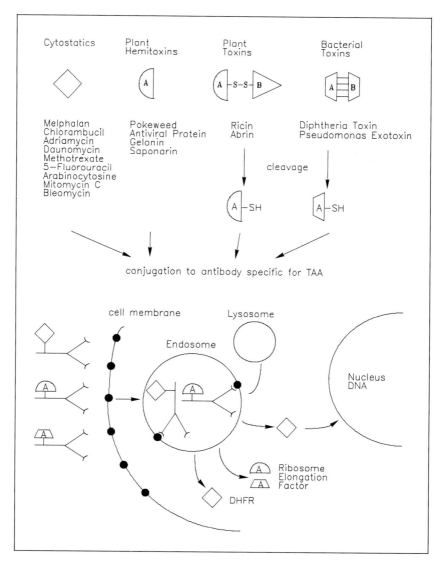

Fig. 13. Schematic demonstration of the cytotoxic activity of cytostatics or toxins and fragments of toxins linked to antibodies specific for TAA.

systems have a higher predictive value for the screening of cytotoxic drugs than animal tumors do [198, 516, 680]. As already discussed (see p. 53 ff.), the localization of specific antibody to the tumor site is much higher

Table XLIV. Experimental chemoimmunotherapy with immunotoxins

Antibody	Toxin	Toxin/antibody ratio	AB binding (%)	Exposure time (h)/test system	Cytotoxicity in vitro activity	Cytotoxicity in vitro specificity	Antitumoral activity in vivo (appl. tumor/drug)	Author
454 A12 (TFR)	ricin A	~1	100	24/PS	> AB	+	> AB (i.p./i.p.)	Fitzgerald et al., 1987
anti Thy 1,2	ricin A	8	70	16/Lc	< ricin	+	> AB (i.p./i.p.)	Blythman et al., 1981
B3/25 (TFR)	ricin A	1–2	100	20/Lc	= ricin	+	> AB (s.c./i.p.)	Trowbridge et al., 1981
	diphth. A	1–2	100		< diphth. A	+	> AB	Trowbridge et al., 1981
7D3 (TFR)	ricin A	1–2	100	6h/PS	> AB; > control conjugate		> AB	Griffin et al., 1987

PS = protein synthesis
Lc = incorporation of ^{3}H-leucine
AB = antibody
TFR = transferrin receptor

(1,000 ×) in xenografted tumors than in human patients. Consequently, successful therapeutic experiments in xenografted tumors sensitive to the cytostatic drug conjugated to the antibody do not seem to be a sufficient prerequisite for predicting the immunospecific therapeutic effect of the conjugate on the tumor for a similarly sensitive tumor in the patients.

Clinical Data and Pharmacokinetic Considerations

In spite of these critical considerations, the therapeutic effect of immunocytostatics and of immunotoxins on experimental tumors as well as the attractive idea of a possible target-specific tumor therapy have been a stimulus to perform clinical studies, so much the more in that all cytostatics used for linkage to antibodies are standard therapeutics in tumor therapy. Indeed, some selected phase-I studies have already been performed (see table XLV).

In a study by Ghose et al. [228] 13 patients with inoperable recurrent malignant melanoma were treated with chlorambucil bound to goat or rabbit antihuman melanoma Ig. Follow-up studies ran for a minimum of 29 months or until death. Two patients showing an objective response to immunochemotherapy had the disease confined to lymph nodes and cutaneous sites; 5 others showed stabilization of cutaneous, nodal, and visceral metastases; and 6 patients showed progression of their disease. The median survival of the responders and stabilizers was 20 months, but only 3.5 months for patients with disease progression. No hematological or renal toxicity was detected after immunochemotherapy.

Ford et al. [203] treated 8 patients with advanced metastatic carcinoma (4 colorectal and 4 ovarian) with a single dose of immunoconjugate increasing from 1.2 to 42 mg antibody (goat anti-CEA) linked to 24–1,800 μg VDS. The in vitro activity of the anti-CEA antibody and its ability to localize in vivo were preserved after conjugation. There was no obvious toxicity or hypersensitivity of the therapeutic effect attributable to either the radiolocalization or escalated doses of conjugate in any of the patients.

A phase-I trial of an immunotoxin constructed with ricin A chain and a monoclonal antibody against melanoma has also been completed in metastatic melanoma. The total doses applied ranged from 0.08 mg/kg daily for 5 days to 1 mg/kg daily for 4 days (total dose: 3.2–300 mg). In 22 patients treated 1 complete response, 4 mixed response and 5 stable disease for more than 2 months were reported. No major problem with toxicity

Table XLV. Clinical trials with immunocytostatics and immunoconjugates

Patients/disease	No.	Treatment	Results	Author
Inoperable recurrent melanoma	13	chlorambucil bound to goat or rabbit anti-melanoma Ig	2 MR ⎤ MS 5 SD ⎦ 20 months 6 PD, MS 3.5 months no hemato- or renal toxicity	Ghose et al., 1977
Melanoma (metastatic)	22	ricin A conjugated to MAb anti-melanoma	transient fall in serum albumin, protein edema, malaise, fatigue, fever, allergic reactions; 1 CR 4 MR 5 SD 11 PD 1 not evaluable	Spitler, 1986 Spitler et al., 1987
Neuroblastoma advanced disease	6	daunorubicin or chlorambucil, conjugated to allogeneic Ig of hyperimmunized volunteers	2 MR 4 alive and disease free	Melino et al., 1982; 1984; 1985
	12		9 'marked anti-tumor response'	
	7		6 PR 1 MR	
Colorectal cancer Ovarian cancer	4	vindesine conjugated sheep anti-CEA antibody	no side effects; no therapeutic effect	Ford et al., 1983

CR = complete response
PR = partial response (tumor reduction > 50%)
MR = mixed response (tumor reduction < 50%)
SD = stable disease
MS = mean survival

Table XLVI. Systemic chemoimmunotherapy with immunoconjugates — dosimetric aspects

	No. of molecules for killing 1 sensitive tumor cell		Toxin/antibody ratio*	No. of conjugated antibodies to kill					
				1 tumor cell		in vitro 10^7-10^9 tumor cells			in vivo (man)** 10^7-10^9 tumor cells (1g tumor)	
						molecules		amount	amount	
Probability of cell deterioration:	50%	99%		50%	99%	50%	99%	50%	99%	50%	99%
ADM	5×10^6***	5×10^7***	10:1	5×10^5†	5×10^6	$5 \times 10^{12} - 5 \times 10^{14}$	$5 \times 10^{13} - 5 \times 10^{15}$	1.25-125 μg	12.5-1,250 μg	8-830 mg	0.08-8 g
Ricin A	1	50 (10-100)	1:1	1	50 (10-100)	$1 \times 10^7 - 1 \times 10^9$	$5 \times 10^8 - 5 \times 10^{10}$	3-300 pg	0.15-15 ng	0.02-2 μg	1-100 μg

Conditions:
* No impairment of antibody epitope reaction
** In vivo distribution of antibody similar to conjugated antibody; localization of antibody 0.015% / g tumor. All tumor cells expose similar amounts of epitopes of similar accessibility
*** Tumor cell is sensitive to ADM
† One tumor cell exposes $\geq 5 \times 10^6$ epitopes

ADM = adriamycin

contraindicating the clinical use of the immunotoxins has been found. Three main toxicities observed in this trial were lethargy and malaise, evening fever, reversible hypoalbuminemia without proteinuria and edema [628, 629].

Melino et al. [454–457] treated a number of neuroblastoma patients with antibody-conjugated drugs. Allogenic polyclonal antineuroblastoma antibodies prepared from the serum of haploidentical volunteers immunized with irradiated neuroblastoma cells were used. Two patients treated with daunorubicin conjugates showed good initial responses, then relapsed and died after 3 and 8 months. Four remained alive and disease-free after 6–19 months of treatment. No drug side effects were observed. They later reported that 9 of 12 patients treated with conjugates of both daunorubicin and chlorambucil (30 mg IgG/kg, twice a week) showed marked 'antitumor responses' with no detectable anti-idiotypic or antiallotypic antibodies or other blocking factors. No toxic side effects were noted. A recent report from this group reported on 7 patients with advanced neuroblastoma treated following a protocol consisting of 1 conjugated chlorambucil (0.5 mg/kg) and 2 conjugated daunorubicin (1 mg/kg) injections per week for 1 year. Partial regression of tumors occurred in patients with stage IV disease, while those with less than stage IV cancer had no evidence of disease after 3 years.

As in any other phase-I study in tumor therapy, these data cannot be taken as proof of the clinical efficacy of immunocytostatics or immunotoxins. Randomized prospective and adequately controlled clinical studies have to be performed subsequently to confirm the above-mentioned preliminary and favorable clinical data. However, it seems reasonable to gather all the knowledge we have so far on the pharmacology and pharmacokinetics of cytostatics, toxins and antibodies and to evaluate whether and under what conditions immunoconjugates might have a chance of being therapeutically effective in tumor patients. This evaluation has been carried out for adriamycin – in comparison to ricin A conjugates (see table XLVI). From our intracellular pharmacokinetic studies (Hoffmann, Kraemer, Sedlacek, unpublished data) we estimated that about 5×10^7 molecules of adriamycin have to be inside an adriamycin-sensitive cell to kill it with a certainty of 99%, in the case of ricin A about 1–100, on average 50 molecules per cell. If adriamycin or ricin A chains are conjugated to a specific antibody with conjugation ratios of about 10:1 or 1:1, respectively, we would need about 5×10^6 molecules of the adriamycin immunoconjugates or about 50 molecules of the ricin A immunoconjugate to destroy one tumor cell. Hereby the tumor cell has to expose at least 5×10^6 binding sites (specific epitopes) for the adriamycin immunoconjugate and 50 binding sites for the ricin A immunoconjugate. In a case where 1 g tumor contains between 10^7 and 10^9 tumor cells and where all these tumor cells expose the corresponding epitope in sufficient amounts, we consequently need about 12.5 μg–1.25 mg of the adriamycin conjugate and 0.15–15 ng of the ricin A conjugate to sterilize all tumor cells within this tumor mass (see table XLVI). In order for this amount of immunoconjugate to be localized to 1 g of tumor mass growing subcutaneously in a mouse, the mouse has to be intravenously injected with about 0.08–8.3 mg of the adriamycin conjugate and 1–100 ng of ricin A conjugate. This calculation is based on the assumption that the respective immunoconjugate localizes to the same degree at the tumor site as is known of the corresponding free antibody. 10–50% (on average about 15%) of the i.v. injected radiolabelled antibody specifically localizes in 1 g of subcutaneously growing tumor xenografts. In man, however, the percentage is in the range of 0.015%/g tumor (see p. 53ff. and p. 68ff.). Thus about 0.08–8 g of adriamycin conjugate and 1–100 μg of the ricin A conjugate may be administered to patients to sterilize 1 g of tumor containing 10^7–10^9 tumor cells. In this calculation we have considered neither the heterogeneity in the expression of TAA within a tumor, nor all the different physiological and anatomical parameters which influence the access of antibodies to the tumor site. It seems doubtful whether the optimistically estimated amount of up to 8 g of adriamycin-conjugated antibodies, when applied to one patient for one course of treatment, can be tolerated by him/her without causing extreme side effects. Another aspect is the logistics of the production of such huge amounts of adriamycin-conjugated antibodies. As demonstrated by ricin A conjugated antibodies, the selection of a cytostatic molecule exhibiting a molar cytotoxicity which is far higher than that of adriamycin for conjugation to an antibody may reduce the amount of immunoconjugate needed to sterilize a tumor. Through the use of such immunoconjugates, tumor cells which expose far less (about 50–100) than 5×10^6 epitopes on their cell membrane might be killed. At first sight, ricin A might be the candidate of choice in fulfilling these requirements. It is, however, extremely immunogenic [272, 629] which makes repetition of treatment difficult or even impossible. On the other hand, the toxicity of the ricin A immunoconjugate is so high that any cell which unspecifically or specifically incorporates only few (1–100) molecules of this immunoconjugate can be destroyed.

It may be suspected that due to this extremely low threshold of cytotoxicity a therapeutic range between unspecific binding in normal cells and specific binding to tumor cells does not exist. In view of the fact that more than 99% of the immunoconjugate does not localize at the tumor site but will be incorporated and/or metabolized by normal tissues, side effects would outweigh any possible antitumoral effect.

In spite of these calculations, the general toxicity of an antimelanoma monoclonal antibody ricin A chain immunotoxin in female Sprague-Dawley rats which received 14 consecutive daily i.v. injections at doses of 1 mg/kg/day or 5 mg/kg/day and were killed at day 22 at the latest, was astonishing low [272].

Moreover, the results of a phase-I study by Spitler et al. [629] with a ricin A melanoma antibody immunoconjugate showed that the side effects were acceptable and objective tumor responses were reported (see table XLV). We have to wait and see whether these data can be reproduced and whether the tumor responses seen in a phase-I study after application of the immunoconjugate can be substantiated in a randomized prospective study.

Nevertheless, future research in immunoconjugates should concentrate on finding new cytostatic agents which are stronger in their cytostatic activity than adriamycin, but weaker than ricin, and which after binding to antibodies exhibit such a degree of cytotoxicity that cells unspecifically incorporating only few immunoconjugates are not affected, whereas accumulation above a threshold, through the antibody specificity of the immunoconjugate, can destroy the cell.

Moreover, just as in the case of radioimmunotherapy, the amount of antibody localizing at the tumor site must be improved by the selection of new antibodies which recognize tumor-selective epitopes on those TAA which are exposed by the tumor cell in high amounts, with significantly increased affinity. Moreover, techniques should be developed to construct antibody fragments small enough to endow them with an increased transcapillary penetration rate but large enough to be conjugated to the cytostatic compound without affecting the immunological function. In addition, new conjugation procedures should be elaborated, which neither affect the antigen-binding activity of the antibody nor the cytotoxic potency of the cytotoxic drug, but drastically reduce unspecific binding of the immunoconjugate to normal tissues.

Even if such immunoconjugates are developed in the future, the problem of target cell restriction of cytotoxic action of immunoconjugates cannot be solved. Due to the principle of action that immunocytostatics or immunotoxins only destroy those cells to which the immunoconjugate sufficiently binds, all other tumor cells which escape binding either by shedding of TAA of by lack of exposure of this specific TAA, are spared and resistant to that treatment. From this point of view, chemoimmunotherapy with immunoconjugates would be inferior to radioimmunotherapy with ß-emitters (see p. 68ff.).

In summary, a great number of techniques have been developed to link cytostatics and toxins from plant and bacteria to antibodies specific for tumor-associated antigens without deteriorating the antigen-binding capacity. In a considerable number of cases these immunoconjugates proved in vitro to be specifically cytotoxic for the tumor cells expressing the corresponding TAA and were mostly superior to the cytotoxicity of the free drug. This superiority was also found in vivo in the therapy of experimental tumors and of human tumor xenografts.

Only a small amount of data is available regarding human patients. Results of phase-I studies indicate some efficacy, but the pharmacokinetic studies cast doubts upon whether tumor therapy with those immunoconjugates developed up to now might be possible. In the case of adriamycin-conjugated antibodies, the molar cytotoxicity seems to be critical. Localization of specific antibodies at the tumor site might not be high enough to carry sufficient adriamycin molecules to the tumor to destroy it. In the case of ricin A conjugates, the molar cytotoxicity is so high that there seems to be no real therapeutic differentiation between unspecific destruction of normal cells and specific destruction of tumor cells. Moreover, ricin A is immunogenic, which makes it difficult to repeat treatment with ricin A conjugated antibodies.

Future research, therefore, has to concentrate on the search for new compounds with a sufficient molar cytotoxicity and an acceptable therapeutic differentiation between unspecific destruction of normal cells and specific destruction of tumor cells. In addition, the quality of the carrier antibody has to be improved, both with respect to specificity as well as avidity. These parameters are decisive for the relative amount of antibody which localizes at the tumor site.

Moreover, antibody fragments should be developed to which cytotoxic compounds can be linked and which can penetrate to and localize at the tumor site much better than IgG or F(ab')$_2$ fragments.

In addition, the development of new conjugation methods should aim to preserve the binding activity of the antibody as well as the cytotoxic potency of the cytotoxic drug, but also to reduce unspecific binding of the immunoconjugate to normal tissue.

Specific Immunotherapy

Mechanisms of Tumor Cell Destruction

There are several ways (see table XLVII and figure 14) in which MAbs may interact with the host immune system to produce a cytotoxic antitumor effect [160, 604].

Complement-mediated cytotoxicity (CMC) involves the binding of the antibody to cells which express the target antigen, followed by the fixation of complement and cell lysis.

Antibody-dependent cell-mediated cytotoxicity (ADCC) is effected by mononuclear effector cells. Their Fc receptors bind them via the Fc part of MAbs to target cells to which the MAbs are attached. Alternatively, the effector cells may attach themselves to circulating antibodies via the Fc receptor and then be transported to the target cell. In either case, it is the final antibody effector cell conjugate which produces cell lysis. Originally it was postulated that only MAbs of IgG_{2a} isotype are able to mediate ADCC [292]. Recently, however, it has been proved that MAbs of other isotypes such as IgG_{2b}, IgG_3 and even IgG_1 may also induce this cytotoxic function [71, 287, 295, 654, 656]. Previous studies suggested that not only macrophages but also cells with 'natural killer activity' (NK cells) are important effector cells in in vivo and in vitro ADCC [3, 4, 292, 295, 654, 656, 672]. Exactly which cell type is mainly responsible for tumor elimination seems to be dependent on the isotype of the MAb and on the corresponding antigen recognized by the antibody-effector cell conjugate [287, 295, 480a, 654]. In animal studies where tumor-bearing nude mice were used, it was noted that ADCC-mediating MAbs clearly proved to be superior in the suppression of tumor growth when compared to MAbs which do not mediate this function [292, 295, 480a, 654, 656]. Furthermore, simultaneous injection of these MAbs together with mononuclear effector cells or injection of effector cells 'armed' with ADCC-mediating antibodies increased tumor destruction [654, 656]. These data encouraged the start of clinical trials with MAbs in oncology.

Table XLVII. Immunotherapy of tumors with monoclonal antibodies — mechanisms of tumor destruction or growth inhibition

— Complement-mediated cytotoxicity (CMC)
    Antibodies
    Antibodies + fresh serum
    Antibodies + mediators
— Antibody-dependent cell-mediated cytotoxicity (ADCC)
    Antibodies
    Antibodies + mediators
    Antibodies + mononuclear cells (NK; M)
    (separate or armed cells)
— Phagocytosis of opsonized circulating tumor cells
— Activation of cytotoxic immune cells
— Antibody-mediated direct cytolysis of tumor cells
— MAb against growth factor receptors
    (IL-2; transferrin; EGF; PDGF)
— Induction of anti-idiotypic response

A third mechanism of antibody-mediated tumor elimination is phagocytosis of opsonized circulating tumor cells by phagocytic cells in the reticuloendothelial (RE) system. A similar process may occur at the tissue level where opsonized cells may attract phagocytic macrophages from the tissue or the circulation [108, 160].

MAbs may also directly activate immune cells, e.g. T cell subpopulations, as was demonstrated for MAbs directed against the ganglioside antigen GD_3 [299], and thus indirectly cause tumor lysis.

Monoclonal antibodies may have an anti-tumor effect via a regulatory mechanism rather than direct cytotoxicity on the target.

One application may be the use of anti-idiotype antibodies as regulatory therapy. Inoculation of human subjects with mouse MAbs results in the production of anti-idiotype antibodies that react with the binding site of the MAb. This reaction is hapten-inhibited, suggesting that an internal image of the antigen is produced by the anti-idiotype response [187, 373]. The possible presence of the internal image of cancer antigen on the human immunoglobulin molecule may change the conditions under which the immune system reacts to the tumor antigen and may open up new approaches for the control of tumor growth. Specific immunotherapy with an appropriate anti-idiotype MAb might provide a critical regulatory feedback effect which restores balance in a particular antigen-antibody system.

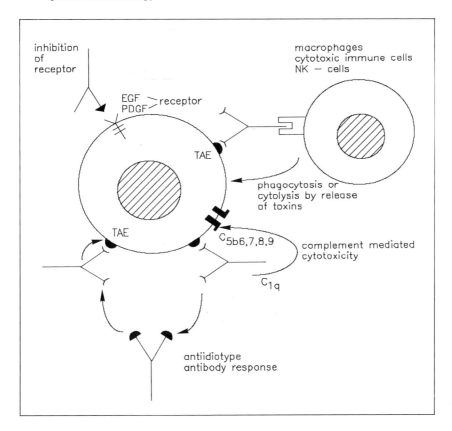

Fig. 14. Mechanisms of antibody-mediated tumor cell destruction.

A second area of regulatory therapy may be the use of MAbs directed against growth factors or growth factor receptors. Some of the growth factors are more essential for activated and proliferating cells than for normal resting cells. For example, MAbs against transferrin, IL-2 receptor, epidermal growth factor receptor and platelet-derived growth factor receptor may be useful therapeutically via regulatory mechanisms [160, 577].

A special approach in tumor therapy could be the application of those MAbs which inhibit functional properties of tumor cells. For example, antibodies directed against ganglioside antigens block tumor cell-substratum interaction and thus prevent invasive and metastatic growth of cancer cells without being directly cytotoxic [125, 126, 163, 510].

A further example is our recent discovery that, contrary to normal cells, pancreatic tumor cells exhibit a superoxide anion production, exocytosis of lysosomal enzymes, pinocytosis and phagocytosis to a degree similar to that known for macrophages. Similar to macrophages, these cell functions of pancreatic tumor cells can be stimulated by antigen-antibody immune complexes as well as by zymosan and can be inhibited by MAbs specifically binding either to pancreatic tumor cells (BW 494/32) or to pancreatic tumor cells and macrophages but that cannot be inhibited by nonspecifically binding MAbs [67, 360, 648, 649].

Treatment of Patients with Leukemia and Lymphoma

The main clinical studies on the treatment of leukemia or lymphoma with monoclonal antibodies are summarized in table XLVIII. In the case of leukemias, infusions of MAbs are accompanied by a rapid fall in circulating leukemia cell levels [155–157, 202, 460–463, 573]. This response is dependent on the MAb used, the rate and dose of infusion, the density of antigen expression, and the presence or absence of circulating antigen or anti-mouse antibodies. This decrease in peripheral target cells, however, is only transient and is similar to that seen with leukopheresis. It appears that circulating target cells are eliminated but that they are rapidly replaced by cells from other organs including bone marrow, lymph nodes and possibly spleen [155–159]. As long as MAb levels persist, the leukemia cell count generally remains depressed, but once the MAb has been cleared, there is a return to pre-treatment levels or higher.

In solid tumor sites direct evidence of cell destruction has been seen [159, 161, 462]. For example, brief objective clinical responses were observed in 4 of 10 patients with CTCL and 2 of 6 CLL patients after therapy with MAb T101 [161]. Acute antitumor effects of T101 were substantially more dramatic in CTCL than CLL but appeared limited by antigenic modulation as observed in other studies on leukemia. In CLL and ALL no long-lasting benefit was seen following MAb therapy [155–157, 159, 202, 573].

Monoclonal antibodies specific for the idiotype of the immunoglobulin produced by the B cell lymphoma and functioning as a tumor-specific antigen of class I (see p. 5ff.) have been used for immunotherapy.

Some dramatic effects of treatment with anti-idiotypic MAbs have been reported on in patients with advanced B cell lymphoma [451, 463]. Patients received IgG_1, IgG_{2a} or IgG_{2b}. One of 13 patients remained in a

Table XLVIII. Clinical trials with unconjugated MAbs or anti-idiotype MAbs in leukemia and lymphoma

Authors	MAb	Disease	Response
Miller et al., 1985	anti-idiotypic	B cell lymphoma	1/13 CR
Meeher et al., 1985			6/12 MR
Garcia et al., 1985	anti-idiotypic	B cell lymphoma	1/10 CR
			5/10 PR
Rankin et al., 1985	anti-idiotypic	B cell lymphoma	2/2 MR
Hamblin et al., 1987	anti-idiotypic (chimeric)	follicular lymphoma	1/1 PR
Bertram et al., 1986	T101	CTCL	1/8 CR
			1/8 PR
Dillmann et al., 1986	T101	CTCL	4/10 MR
Dillmann et al., 1986	T101	CLL	2/6 MR
Miller et al., 1983	anti-T-cell	CTCL	5/7 MR
Ritz et al., 1981	anti-CALLA	ALL	no response

CR = complete response
PR = partial response (tumor reduction > 50%)
MR = mixed response (tumor reduction < 50%)
CTCL = chronic T cell leukemia
CLL = chronic lymphatic leukemia
ALL = acute lymphatic leukemia
CALLA = common acute lymphatic leukemia antigen

complete remission 50 months after receiving the antibody. Six of 12 additional patients have had objective remissions which were also clinically significant [451, 463]. However, these remissions were not complete and were of relatively short duration.

In other studies using anti-idiotypic MAbs in 'low grade' B cell lymphomas, partial remissions were observed [268, 269, 552]. However, the striking effects demonstrated in the first patient treated in Stanford could not be repeated. It is interesting that variations in the characteristics of the individual tumors such as antigen sites per cell and ability to modulate the surface immunoglobulin were not predictive of response. Moreover, therapeutic outcome correlated to the number of host non-tumor cells infiltrating the tumor. The vast majority of these non-tumor cells were mature T-lymphocytes of T_3 and T_4 phenotype. Thus, a preexisting host-tumor interaction seems to be important in the in vivo effect of anti-idiotype antibodies in B cell tumors [422, 423].

The limited therapeutic success of anti-idiotypic antibodies in B cell lymphoma is mainly due to spontaneous generation of somatic variants not expressing the specific idiotype [268, 269, 278, 461, 622].

Treatment of Patients with Solid Tumors

The therapeutic use of cytotoxic MAbs in early phase I and II trials has recently been evaluated in gastrointestinal tumors, tumors of neuroectodermal origin such as melanoma and neuroblastoma and in lung cancer (see table XLIX).

Table XLIX. Clinical trials with cytotoxic MAbs alone or in combination with effector cells in patients with solid tumors

Author	MAb	Disease	Response
Oldham et al., 1984	9.2.27	melanoma	no
Goodman et al., 1986	96.5	melanoma	no
	48.7		
Houghton et al., 1986	R 24	melanoma	25% PR or SD
Dippold et al., 1985	R 24	melanoma	20% MR
Cheung et al., 1986	anti GD$_2$	neuroblastoma	20% PR or SD
Sears et al., 1985	CO 17-1A + BC	GI adeno ca.	5% PR
			10% SD
Sindelar et al., 1986	CO 17-1A	pancreatic ca.	20% MR
Tempero et al., 1986	CO 17-1A	pancreatic ca.	15% SD
Douillard et al., 1986	CO 17-1A + BC	colorectal ca.	3% CR
			5% MR
			25% SD
Frödin et al., 1986	CO 17-1A + BC	colorectal ca.	25% MR
			50% SD
Schulz et al., 1987	BW 494/32	pancreatic ca.	8% MR
	(BI 51.011)		38% SD
Timms et al., 1986	KS 1/4	lung ca.	no

CR = complete response
PR = partial response (tumor reduction > 50%)
MR = mixed response (tumor reduction < 50%)
SD = stable disease
BC = buffy coat cells
GI = gastrointestinal
Ca. = carcinoma

In melanoma, MAbs directed against glycoproteins or gangliosides have been tested clinically. It was demonstrated that MAb 9.2.27 directed against a 250,000 dalton chondroitin sulfate proteoglycan core glycoprotein [94, 95] selectively localized in vivo to melanoma tissues – however, without inducing any tumor regression [511]. Similar data were obtained when MAbs 96.5 and 48.7 directed against a melanoma-associated surface glycoprotein and proteoglycan antigen were applied to melanoma patients [248]. More successful were clinical trials with the MAb R24 directed against GD_3, a prominent glanglioside on the surface of melanoma cells and other cells of neuroectodermal origin.

The application of this MAb which mediates strong ADCC and CMC and activates T-cell subpopulations [299, 313] induced inflammatory cutaneous responses around tumor nodules and lymphocyte and mast cell infiltration in tumor tissues [164].

In addition, major tumor regressions have been observed in 3 of 12 treated patients [313]. Meanwhile, combination therapy of this MAb with IL2 has been tested. Similar data were obtained when an anti-GD_2 MAb (IgG_3) was tested in neuroblastoma patients [127]. However, the application of this immunoglobulin induced severe pain in the neural system.

In gastrointestinal cancer the most common MAb used therapeutically is CO 17-1A. This antibody was raised at the Wistar Institute and is directed against a cell-surface glycoprotein mainly expressed on colorectal and pancreatic cancer [584]. In a series of investigations it was demonstrated that MAb CO 17-1A mediates ADCC by activating macrophages and is efficient in tumor suppression in nude mice [291, 292, 671, 672].

Single doses of this antibody up to 1,000 mg were well tolerated in patients. The most common side effect was mild gastrointestinal symptoms [421]. However, after repeated doses, over 80% of patients developed human antibodies to 17-1A within 15 days, resulting in allergic responses [421, 597–599].

In patients with pancreatic cancer, response rates to treatment with CO 17-1A either applied alone or in combination with mononuclear cells ranged from 10–20%; however, most responses were only minor tumor regressions [621, 685]. In a large study on 95 patients suffering mainly from colorectal cancer, 3 complete responses, 5 partial responses inferior to 50% and 24 stable diseases were noticed [168]. Other papers have described similar response rates in these tumors [214].

Another MAb directed against a carbohydrate epitope located on a >200 k daltons glycoprotein mainly expressed on well-differentiated adenocarcinomas of the pancreas (BMA 494/32, see table L) has been tested clinically in phase-I/II trials [71, 657].

In a multicenter study on 39 patients with advanced pancreatic cancer, daily doses over 10 days (highest single dose 180 mg, highest cumulative dose 490 mg) were well tolerated [657]. Similar to the experience with CO 17-1A, all patients developed antimurine antibodies between day 14 and 21 (see table LI) including anti-idiotypic antibodies [73]. Prolongation of injections over day 14 resulted in allergic reactions in patients. The response rate was similar to that in the clinical trials with CO 17-1A: Out of a total of 27 patients eligible for evaluation of tumor response (see table LII) 2 were present with objective tumor regression documented by CT scan, and 10 patients responded with a long period of stable disease of up to 40 weeks [657].

Table L. Characterization of BW 494/32 (Bosslet et al., 1987; Schorlemmer et al., 1988)

Isotype:	IgG_1
(Switch mutant:	IgG_{2a})
Specificity:	pancreatic carcinoma grade I, II, ductuli of pancreas
Epitope:	carbohydrate on a > 200 KD glycoprotein of mucin type
Effector functions:	CML 0
	ADCC ++
Effect on pancreatic carcinoma cell function:	inhibition of O_2 radical formation
	inhibition of exocytosis
	inhibition of pinocytosis
	inhibition of phagocytosis
	inhibition of growth (xenotransplanted pancreatic tumor)

Table LI. Human antibodies against murine antibodies (HAMA)

Antibody	Dose	No. tested	No. HAMA positive	No. side reaction
BW 431/31-^{111}In* F(ab')$_2$	1 x 1 mg	58	6 (10%)	0
	2 x 1 mg	16	8 (50%)	0
BW 494/32** (IgG_1)	(14 x Q2) starting dose 100 mg, cumulative 490 mg	18	17 (94%)	3

* Baum et al., 1987
** Schulz et al., 1987

Table LII. Phase-I clinical study in advanced pancreatic cancer (Schulz et al., 1987)

MAb:	BW 494/32
Dose and schedule:	highest single dose 180 mg; highest cumulative dose 490 mg; repeated daily doses over 10 days
Eligible for evaluation:	27
MR (objected by CT scan):	2
SD (up to 40 weeks):	10

MR = minimal response
SD = stable disease

Randomized phase-II trials have started in patients with a relatively small tumor burden to further evaluate the clinical efficacy of this antibody.

In lung cancer patients a phase-I/II trial with MAb KS 1/4, which was developed at R. Reisfeld's laboratory, at the Scripps Clinic, La Jolla, showed good tolerability; however, no major tumor responses have been observed with the application of the antibody alone. Conjugates of this antibody with methotrexate and vinblastine are being tested in animal studies in order to prepare further clinical trials [96].

Immune Response Against Murine Immunoglobulins

The administration of murine monoclonal antibodies to patients results in the development of human anti-murine immunoglobulin response [159, 202, 296, 463, 511, 531, 597, 598, 612]. The degree to which the human recipient responds against murine antibodies (human antibody against murine antibody = HAMA) depends on its amount: in the tumor imaging studies by Baum et al. [38] a single application of 1 mg antibody caused HAMAs in about 10% of the patients. After repeating this dose the percentage increased to about 50% (see table LI). The dose of 5 mg/day seems to induce a strong sensitization [124, 330]. Dosages of 100 mg and more of MAbs given in therapeutic studies caused HAMAs in nearly all patients (see table LI) and induced side effects in some of those patients, in those cases where the dose was applied repeatedly [657]. These side effects were reported to be moderate, compared to the ones caused by the application of polyclonal antibodies [124, 515]. HAMAs may be induced additionally to a preexisting anti-murine immunoglobulin reactivity. This preexisting anti-murine immunoglobulin reactivity has been found in the sera of healthy persons, tumor patients as well as patients with rheumatoid arthritis, and it is due to IgM polyclonal rheumatoid factors which recognize protein epitopes on the Fc part (C-2 domain) of the murine and to a greater extend of the human IgG [144-146, 651, 690].

The antimurine immunoglobulin response may limit the clinical application of monoclonal antibodies, in part by causing allergic reactions when treatment is repeated. On the other hand, the consequence of immunization is an accelerated clearance of the injected MAb due to immune complex formation. This enhanced clearance may significantly inhibit immunospecific localization and efficacy of the injected MAb. Two types of specificity pattern of the antibody response to monoclonal murine immunoglobulins have been identified: those directed against the isotype and its subtypes and/or antibodies directed against the idiotype [124, 187, 296, 330, 373, 612].

In our own studies we were able to differentiate the specificity pattern of the antibody response of pancreatic tumor patients against the murine immunoglobulin BW 494/32 by absorption of sera (table LIII) as well as of supernatants of EBV-infected and hybridized (SP-2) peripheral lymphocytes (table LIV) of those patients [72]. Sera as well as supernatants of lymphocytes were able to block the specific binding of the MAb BW 494/32 to its target cell completely, irrespective of which isotype (IgG_1 or a switch mutant IgG_{2a}) was used.

Thus, according to the classification of Jerne [335], Jerne et al. [334], Kohler [371] and Bona and Kohler [61], we are having to deal with a β- and/or a γ-type of anti-idiotype antibody, the variable region of which either completely or in part (γ-type) mimicks the structure of the epitope to which the murine MAb BW 494/32 binds. According to the idiotype network theory, the patient should also develop antibodies which bind to the paratope of the anti-idiotype antibody. In case this second (human) antibody (antibody 3) is of β-type, its variable region should represent the same paratope as the original murine monoclonal antibody.

As the anti-idiotype antibody of the β-type or an adequate mixture of anti-idiotype antibodies of the γ-type, expresses an internal image (that is by definition the epitope of the TAA) immunization with such anti-idiotype antibodies should induce specific responses against the corresponding epitope of the TAA. Indeed, anti-idiotype antibodies have already been

Table LIII. Evaluation of human anti-murine IgG antibody (HAMA) — response against BW 494/32 (IgG$_1$) (Bosslet et al., 1987)

Patient No.	HAMA titer in serum	% reactivity after absorption of HAMA with (10 µg of)			Valuation
		BW 494/32 IgG$_1$	BW 494/32 IgG$_{2a}$	BM 431/31 IgG$_1$	
1	16	37	62	64	anti-idiotypic anti-isotypic
2	16	34	48	83	anti-idiotypic
3	32	30	40	96	anti-idiotypic

Table LIV. Evaluation of the clonal frequency of anti-idiotypic or anti-isotypic antibody secreting B cells in patients developing HAMA to BW 494/32 (Bosslet et al., 1987)

Patient No.	Growing clones/ 10^5 PBM	No. of hybrids secreting human IgG binding to MAb			Valuation	
		BW 494/32 IgG$_1$	BW 494/32 IgG$_{2a}$	BW 431/31 IgG$_1$	anti-idiotype %	anti-isotype %
1	2.7	31	14	17	45	55
2	4.6	34	33	1	100	0
3	1.1	33	33	1	97	3
4	43.0	3	3	0	100	0

used as vaccines to protect against bacterial, viral and parasitic infections [356, 357, 448, 557, 591, 666]. Anti-idiotype antibodies have also been used in the treatment of B cell lymphoma (see table XLVIII), but in many cases the clinical success was limited due to spontaneous generation of somatic variants [268, 269, 278, 312, 461, 570, 622]. Experimental studies in various solid tumor models show that through the application of anti-idiotype antibodies one can obtain tumor-specific immunization [204, 298, 405, 406, 494, 556]. Since the regression of solid tumors in patients receiving murine MAb specific for TAA turned out to be a late event rather than an early one in nearly all cases, and since it was accompanied by the occurrence of anti-idiotype antibodies (signalizing stimulation of the anti-idiotypic network), an association between these two occurrences may be assumed.

Consequently, anti-idiotype antibodies may clinically be tested as an active specific therapy for solid tumors. This kind of treatment, however, carries the risk that resistant tumor cells negative for the corresponding TAA might be the ones selected, as in the case of direct application of murine MAbs specific for epitopes on TAA or in the case of anti-idiotype antibodies directed against B cell lymphoma cells.

To summarize, up to now several antibody-mediated cytotoxic and antiproliferative mechanisms are known by which monoclonal antibodies might be therapeutically effective in tumor diseases. Indeed, phase-I and phase-II clinical studies in leukemia and various solid tumor diseases have given strong indications of a positive therapeutic effect on tumors. Clinical effectiveness, however, has to be proven by adequately randomized prospective studies. Such studies are being carried out now for certain diseases such as the minimal residual disease of pancreatic carcinoma after Whipple operation [657]. Considering the altogether disappointing outcome of the numerous randomized clinical studies in the therapy of tumor diseases with chemoimmunotherapeutics and immunostimulators [657], tumor therapy using monoclonal antibodies can only be accepted as an effective routine clinical procedure when its activity has reproducibly been proven by such randomized prospective studies. It is already obvious that immunotherapy with monoclonal antibodies in its present form is of limited activity. In fact, dosimetric studies have shown that on average only 0.015% of the injected antibody localizes per g tumor (see p. 53 ff. and p. 68 ff.).

In the case of radioimmunotherapy and chemoimmunotherapy, side effects induced by the residual 99.985% of the injected antibody (conjugated with a cytotoxic moiety) limit the total amount that can be given per patient. However, dose limitation due to side effects is not as critical in immunotherapy. High doses up to 1–2 g per patient can be given, so that localization in amounts of mg of monoclonal antibodies per g tumor can be achieved. Since only antibodies immunospecifically binding to its target activate effector mechanisms, such cytotoxic reactions may be much more concentrated at the tumor site than is the case with radioimmunotherapy and chemoimmunotherapy.

Effector mechanisms can be improved by the selection of a highly effective isotype (or subclass) of the murine monoclonal antibody. In the case of an approved tumor specificity, the isotype subclass may even be improved by isotype switch mutant selection (see p. 31 ff.). In addition, increased expression of TAA by tumor cells and non-antigen-specific stimulation of effector cells (NK cells and macrophages) may be induced by the application of α-interferon, γ-interferon, IL-2 or GM-CSF [252, 253,

254, 730]. The application of immunomediators in combination with monoclonal antibodies may result in better tumor response rates.

Nevertheless, it can already be foreseen that due to the extreme heterogeneity in the expression of TAA (see p. 5ff.) immunotherapy of tumors using monoclonal antibodies might be hampered by the selection of TAA-negative cells, that is, tumor cells resistant to tumor therapy. This problem might be reduced by the application of a mixture of monoclonal antibodies [116]. On the other hand, repeated treatment of patients with murine monoclonal antibodies is limited due to allergic reactions induced by the xenogeneity of the antibody. Anti-isotype and anti-idiotype antibodies are induced.

The clinical course of tumor regressions (they do not occur immediately but with a delay after application of murine monoclonal antibodies), and the frequent detection of anti-idiotypic antibodies in these patients indicates that the generation of anti-idiotypic antibodies may be an essential mechanism of tumor regression. If so, the application of murine MAbs would induce via the idiotype-anti-idiotype antibody network system [334, 335] antibodies within the patient directed against the corresponding epitope on the TAA. These patients' own antibodies would then be effective on the tumor.

The anti-anti-idiotypic reaction may offer us an internal specific amplification system for the antibody response against autochthonous tumors and by specific stimulation would enable us to increase the level of specific antibodies to a degree sufficient for effective tumor regression. On the other hand, it lays open to us a new way of active specific tumor immunotherapy in a sense of 'vaccination'. In experimental systems, this tumor-specific idiotype vaccine has already been shown to be effective [405, 556].

Bone Marrow Purging of Neoplastic Cells

Autologous bone marrow transplantation has recently been introduced in the treatment of leukemia and solid tumors in order to intensify combined chemo- and radiotherapy without damaging bone marrow stem cells.

In patients with disseminated malignancy, however, autologous bone marrow transplantation may include reinfusion of tumor cells which have infiltrated their bone marrow.

Consequently, autologous bone marrow grafts have to be purged in vitro of tumor cells. Methods that have been used to remove tumor cells from marrow (see table LV) include physical separation, cytotoxic drugs, TAA-specific MAbs together with complement, MAb linked to intact ricin or to ricin A chain or to daunomycin, combinations of MAbs against TAA with cyclophosphamide or MAbs linked with pokeweed antiviral protein combined with cyclophosphamide. Any valuation of these methods is difficult, because clearly designed and adequately controlled studies do not exist. Nevertheless, immunotoxins seem to be encumbered with the long time they need (many hours or days) to destroy the target cells [676]. Mixtures of antibodies and complement tend to be toxic to non-neoplastic cells in the marrow and cause clumping of cells [35] and many MAbs require very high antibody or complement concentration to destroy tumor cells [571].

Nevertheless, the purging of bone marrow by MAbs reacting specifically with tumor cells or leukemic cells but not with stem cells or other precursor cells of the hematopoietic system seem to be superior to all other non-target-specific methods to circumvent the problem of toxic damage to bone marrow stem cells. Due to the heterogeneity in the expression of TAA by tumor cells, exposure to a 'cocktail of antibodies' or immunotoxins has been demonstrated to be more effective in eliminating neoplastic cells than the use of a single anti-TAA antibody [35, 36, 82, 564, 676, 707–709].

Efficiency rates of destroying tumor cells higher than 99.9% have been achieved. The best in vitro results were obtained with antibody linked to pokeweed antiviral protein and in combination with mafosamide. This combination eliminated about 7 logs of tumor target cells with minimal toxicity to normal bone marrow stem cells [708, 709].

The first clinical results of studies using MAbs and complement or MAbs conjugated to either magnetic immunobeads or toxins for bone marrow purging are now available [527, 565, 632].

In acute lymphoblastic leukemia good results were obtained using a cocktail of 3 IgM MAbs called 'VIB-Pool' [88, 562, 632]. These MAbs mediated complement-mediated cytolysis (CMC) with human complement; therefore, no rabbit complement is necessary, but the patients' own serum can be used as a complement tool. In patients with CALLA-positive leukemia ('high-risk' or relapsed ALL) it was found that this procedure is safe and no reduced rate of engraftments could be observed [88, 562]. However, the final proof that the number of long-term survivors after this treatment was increased, compared to unpurged transplanted survivors, is still being awaited.

In leukemia, other MAbs reactive with null-cell-type ALL [471], myeloid differentiation antigens present on acute myelogenous leukemia (AML) [28] or T cell antigens present on T cell leukemia have been successfully used for purging.

In solid tumors, preliminary experience with bone marrow purging and autologous transplantation is mainly to be found in patients with neuroblastoma. Clinical trials in small cell lung carcinoma (SCLC) will probably start shortly, as selective MAbs are availabe [509, 662].

A multicenter study (CCG) has been designed for patients with advanced neuroblastoma, to evaluate the efficacy of purging with MAbs conjugated to magnetic immunobeads [605a]. Using a panel of 3 MAbs, it was shown that a combination of sedimentation, filtration and MAbs/magnetic beads removes 5–6 log tumor cells from infiltrated bone marrow [565]. In the case of 17 patients where transplantation was carried out before progressive disease, the estimated survival time was 57% at 46 months, which was better than the 18% survival of a comparable group. However, a final evaluation of this study is not yet available [605a].

In another large trial including 65 stage IV neuroblastoma patients high-dose chemo- and radiotherapy followed by autologous bone marrow transplantation purged by MAbs and immunobeads achieved a better overall survival 27 months after treatment, compared to a historic control group treated by conventional therapy (0.24 versus 0.08, $p < 0.05$) [48].

Further studies have to prove whether or not the above-mentioned therapeutic regimens are really beneficial for patients with malignancies.

Table LV. Bone marrow purging of neoplastic cells

Methods	Authors
Physical separation	
MAbs conjugated to microspheres including magnetobeads	Treleavan et al., 1984
	Seeger et al., 1987
	Reynolds et al., 1986
	Bernard et al., 1987
Cytotoxic drugs	
Cyclophosphamide (activated)	Kaizer et al., 1982
	Ramsay et al., 1985
Anti-TAA antibodies	
MAbs + complement	Ritz and Schlossmann, 1982
	Buckman et al., 1982
	Stepan et al., 1984
	Jansen et al., 1984
	Bast et al., 1985
	Ritz et al., 1982
	Ramsay et al., 1985
	Reisfeld et al., 1986
	Okabe et al., 1985
	Stahel et al., 1985
	Brun del Re et al., 1985
	Reiners et al., 1987
	Morishima et al., 1985
	Ball et al., 1986
	Herve et al., 1986
Antibody conjugates	
MAb – ricin	Thorpe et al., 1982
	Leonhard et al., 1985
	Stong et al., 1984
MAb – ricin A	Muirhead et al., 1983
	Raso et al., 1982
	Krolick et al., 1982
	Myers et al., 1984
	Gorin et al., 1985
MAb – pokeweed anti-viral protein	Uckun et al., 1985
MAb – daunomycin	Diener et al., 1986

Table LV. (continued)

Methods		Authors
Combinations		
MAbs	+ cyclophosphamide (active)	De Fabritis et al., 1985
MAbs	+ mafosamide	Uckun et al., 1985
MAbs	– pokeweed antiviral protein	Uckun et al., 1985
	+ cyclophosphamide	
MAbs	– ricin A	Stong et al., 1985
	+ lactose	
MAbs	– pokeweed antiviral protein	Uckun et al., 1985
	+ ammonium chloride	
MAbs	– ricin A	Bregni et al., 1986
	+ chloroquine	Donay et al., 1985

+ = in combination with
– = conjugated to

To summarize, autologous bone marrow transplantation in heavily irradiated or chemotherapeutically treated tumor patients carries with it the risk of tumor recurrence caused by tumor cells within the bone marrow graft. Several methods have been developed for bone marrow purging of tumor cells without harming normal bone marrow stem cells. Up to now the most effective methods have included a battery of monoclonal antibodies directed against various TAA exposed on the surface of the tumor cells in the bone marrow. Such monoclonal antibodies provide the degree of tumor cell specificity needed in order not to damage the normal stem cells.

Any final judgement about the success of bone marrow purging in autologous bone marrow grafting is not possible, because adequately controlled clinical studies still do not exist. Efforts are still being made to find the most effective procedure of purging, whereby efficacy is calculated according to the degree of log tumor cell destruction without damage to normal stem cells. The more tumor cells are specifically killed the lower is the risk of transplanting tumor cells by bone marrow grafting.

Bone marrow purging is the ideal in vitro assay for proving the effectiveness of anti-TAA MAbs. This system is not hampered by the pharmacokinetics of antibodies in vivo. All other parameters which are of the utmost relevance in vivo are also relevant in this situation, such as heterogeneity and variability in the expression of TAA by tumor cells, the specificity of the antibody, i.e. the extent of its reactivity with hematopoietic progenitor cells, and the cytotoxic potency of the antibody in combination with complement or conjugated with toxins or cytostatics.

In consequence, the final outcome of clinical studies in which autologous bone marrow purged with monoclonal antibodies has been grafted will give us useful information about the value of monoclonal antibodies in tumor therapy. This has to be considered irrespective of the question of whether autologous bone marrow grafting will go on to become an important therapeutical procedure in tumor therapy.

Human or Humanized Monoclonal Antibodies

Human Monoclonal Antibodies

In contrast to the spectacular results of rodent monoclonal antibody technology which has led to successful achievements in diagnosis and to justified hopes of future therapeutical procedures in human cancer disease, the development of human monoclonal antibodies is still at a very early stage. The reasons are manifold. One of the main obstacles seems to be the insufficiency of procedures known to date regarding the immunization of human lymphocytes in vitro [166, 331]. Thus, lymphoid cells from the blood, from tumor draining lymph nodes or from lymphoid infiltrates of tumor tissue have been taken from patients, on the assumption that they might have a specific immune response against their tumors. In some cases, patients have been actively and specifically immunized – e.g. in tumor diseases of the bladder, brain, stomach, colon, breast, lung, kidney, and in melanoma [63, 109, 140–143, 302–305, 326, 484, 523, 618–620, 678, 725, 728, 740]. In spite of these sources, the development of human monoclonal antibodies was hampered, due to deficient methods of selecting specifically committed lymphocytes and recognizing their ideal differentiation stage for fusing them [659] with myeloma cells.

Moreover, an insufficient number of antigen-specific B cells [308, 512, 513], their transient appearance after immunization and their low mitotic activity may have contributed to the altogether negative outcome in generating human hybrids for the production of monoclonal antibodies. In addition, the function of the parenteral cells as well as the resulting hybrids, may have been impaired by the existence of too many cytotoxic T cells and/or suppressor B cells in the various lymphocyte preparations [140].

B lymphocytes can be transformed to continuously growing cell lines by infection with Epstein Barr virus (EBV), via the C3d complement receptor [207]. However, transient activation of B lymphocytes without subsequent stable transformation and/or a transient antibody production for a period of only 1–2 months are major problems [331] of this technique.

Another very important handicap in the development of human monoclonal antibodies is the lack of human myeloma cell lines exhibiting similar qualities to the corresponding murine myelomas.

Human myeloma cell lines grow very poorly in culture [331]. As an alternative, lymphoblastoid cell lines derived from EBV transformation of lymphocytes have been used. These lines are easier to handle in culture than myelomas, but they present problems due to their low level of antibody secretion [376-380]. Improvements have been made by fusing lymphoblastoid cell lines with mouse or human myelomas. These heteromyelomas have been claimed to be superior to mouse or human myelomas in their fusing potency with human lymphocytes [206, 686]. However, heterofusion of human lymphocytes with mouse myelomas has also produced encouraging results [690, 740]. Genetic instability of these heterohybrids does not seem to be worse than that of human hybrids [690], while the fusion rates are significantly higher [142, 143].

Significant improvements have been made by fusing EBV transformed human lymphocytes with human or murine myeloma cells [377]. Advantages of this method are the significant increase in the fusion rate and the higher levels of specific antibodies produced. The problem of instability of these clones can be solved by selecting stable clones.

In our pursuit of human monoclonal antibodies directed to membrane-associated TAA, we ended up with a modification of the technique introduced by Kozbor et al. [377]. Human peripheral blood lymphocytes of lymphoid cells from lymph nodes draining the tumor area are isolated, infected with EBV and cloned. The supernatants of these lymphoblastoid cell lines are checked for antibodies specific for TAA located in the cell membrane. Positive cell lines are fused with murine myeloma SP-2 and subsequently cloned and retested.

Our results are positive in that stable heterohybrids were developed which produced antibodies (mostly of IgM, less of IgG class) in sufficient amounts. However, after the hybridization of lymphocytes taken from peripheral blood as well as from tumor draining lymph nodes, it proved extremely difficult to find hydridomas which secreted antibodies which reacted sufficiently with membrane-associated TAA. The overwhelming majority of antibodies reacted with antigens localized within the tumor cells.

From the data collected from more than 50 tumor patients we can conclude that the occurrence of B cells committed for TAA located at or in the membrane of autochthonous tumors is indeed a very rare observation in tumor patients.

Future research for generating human monoclonal antibodies directed against membrane-located TAA, consequently has to concentrate on methods and procedures to select patients with an established specific humoral immune response against TAA located in the cell membrane.

The number of such patients seems to be extremely low. On the other hand, the identification of a monoclonal antibody specific for a tumor-selective epitope on TAA also is a very rare event.

Consequently, (with the use of peripheral blood lymphocytes or tumor-draining lymph node cells from the tumor patients) the generation of human monoclonal antibodies specific for a tumor-selective epitope on TAA has been and will be extremely rare (see table LVI). An increase in frequency can only be expected when new techniques are developed and effective in vitro immunization procedures are available, or when patients can undergo specific immunization by an adequate preparation which is effective in tumor therapy.

In the case of there being any future success in the development of human monoclonal antibodies directed against tumor-selective epitopes on TAA, the question will still remain as to whether human immunoglobulins, produced by heterohybridomas, are identical to polyclonal human immunoglobulins. A significantly altered antigenicity of human monoclonal IgG, less of IgM and IgA, derived from human-mouse heterohybridomas has been found in serological assays [347], which may also imply an altered immunogenicity of those preparations when applied to patients. However, there might also be problems in the test system, mimicking an altered antigenicity.

Chimeric Antibodies and 'Humanization'

The use of mouse monoclonal antibodies (MAbs) for therapy of human cancer induces an immune response to the foreign protein and prevents continuous application of the mouse MAbs [414] (see p. 109 ff.). The reduction or abolition of this human anti-mouse immune response is an essential predisposition for successful therapy and a long-term diagnostic follow-up with mouse MAbs.

In the past few years recombinant DNA technology has been used to address the problem of antigenicity of mouse MAbs. The basic idea was to reduce the murine nature of MAbs without altering their antigen-binding capacity. In a first attempt the constant (C) regions of the mouse MAb were

Table LVI. Human monoclonal antibodies against tumors

Tumor	Isotype	Technique/ fusion partner	Stability (months)	Author
Melanoma	IgM + IgG	EBV	—	Watson et al., 1983
Melanoma	IgM + IgG	NS-1	—	Abrams et al., 1984
Stomach cancer	IgM	EBV		Hirohashi et al., 1982b
Colorectal cancer	IgM	NS-1	> 12	Haspel et al., 1985
Colorectal cancer	IgM	LICR-LOW-Hmy2	> 12	Borup-Christensen et al., 1986
Breast ca.		NS-1	> 6	Wunderlich et al., 1981
Breast ca.		NS-1 Hmy2		Cote et al., 1983; 1984; 1986
Lung tumor	IgM	Namalwa	—	Murakami et al., 1985
Lung tumor		P$_3$X63-Ag8.653	> 63	Hirohashi et al., 1986
Lung tumor	IgM	EBV	—	Hirohashi et al., 1982
Lung tumor	IgM	EBV	> 12	Cole et al., 1984
Transitional-cell ca.	IgM	EBV	> 1	Paulie et al., 1984
Glioma		LICR-LION-Hmy2	—	Sikora et al., 1982; 1983
Leukemia		RH-L4	—	Olsson et al., 1984
T-cell leukemia	IgG$_1$	EBV	> 2	Matsushita et al., 1986
Lymphocytic leukemia		NS-1	—	Abrams et al., 1984

Ca. = carcinoma
EBV = Epstein Barr virus transformation

replaced with the corresponding human equivalents by a recombination of the murine and human genes of cDNAs [76, 87, 420, 495, 592, 634] (fig. 15).

The antibody molecules obtained by this approach retain their specificity for antigen and thus their usefulness for targeting but are expected to be less immunogenic in man since their constant parts are of human origin. This expectation may not prove true. A more detailed analysis of the human anti-mouse antibody (HAMA) response of 18 patients treated repeatedly with the mouse MAb BW 494/32 ($\gamma_1\kappa$) which is

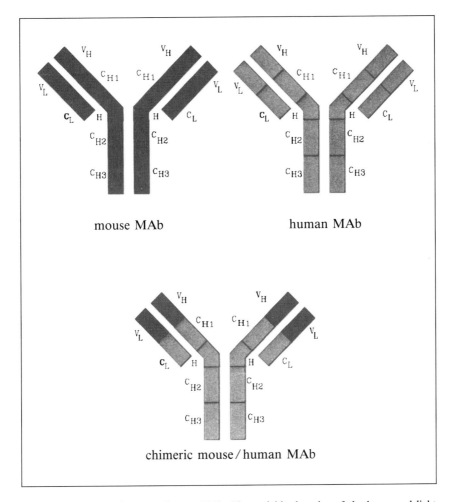

Fig. 15. Chimeric mouse/human MAb: The variable domains of the heavy and light chains of a mouse MAb are combined with the constant parts of a human MAb.

specific for pancreatic, colorectal and stomach carcinomas [71] revealed that in 17 cases the major part of the HAMA response is directed to the Fab' portion of the mouse MAb [73, 657].

In a representative experiment, shown in fig. 16, the binding of HAMA from a patient's serum to MAb BW 494/32 was inhibited by BW 494/32 and to the same extent by its Fab' fragment. No inhibition could be achieved with other immunoglobulins. It is unlikely that the anti-mouse Fab' anti-

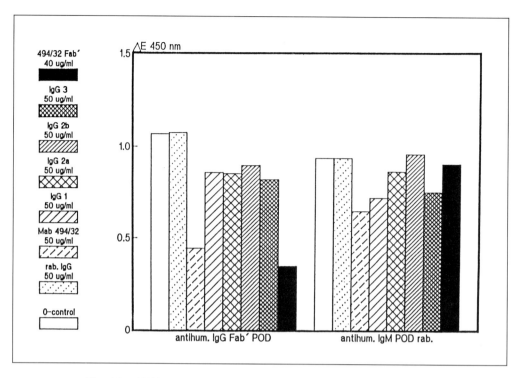

Fig. 16. Inhibition of human antimurine antibodies by different murine immunoglobulins. Serum samples of patients were incubated with murine MAbs of different isotypes and tested in a HAMA assay as described in [8]. In the experiments listed on the left side an anti-human-IgG secondary antibody was used, whereas the data of experiments on the right side of this figure were obtained by an anti-human-IgM secondary antibody.

bodies are directed exclusively at the C-domains; therefore, these results indicate that the HAMA response is at least in part anti-idiotypic. These results were confirmed by a clonal analysis of the HAMA response in several patients. EBV-transformed B cell clones exclusively produce MAbs which are specific for the variable domain of BW 494/32 (see tables LIII, LIV).

The chimeric MAbs which retain the complete mouse variable domains will probably not be able to avoid the HAMA response completely, though none of the existing chimeric MAbs has yet been sufficiently tested in patients to prove this theory. However, according to preliminary information, the use of chimeric immunoglobulins does not seem to prevent the induction of HAMAs [124].

An advanced molecular biological approach was recently successful. It reduces the mouse portion of the MAb to the minimum essential part. Only the antigen-binding loops of the V domains, coded by the complementary determining regions (CDRs) of the V genes, are of mouse origin. The rest of the antibody molecule is human [341, 568].

This technique, called 'humanization', is based on the discovery that the special structure of mouse and human V domains of heavy and light chains, as determined by X-ray chrystallography, are very similar in the

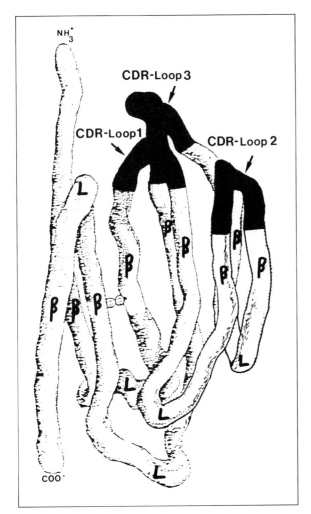

Fig. 17. Schematic drawing of the 3-D structure of an Ig V domain. The black regions of the molecule represent the CDR loops. (From Poljak et al. 1973; modified.)

framework regions. They show considerable differences only in the antigen binding CDR loops [8]. For 'humanization' the CDR loops of a specific MAb are transplanted to the V domains of a human myeloma protein, supplying the human myeloma protein with the antigen-binding property of the specific MAb (fig. 17).

The 'humanization' technique has been applied successfully to 3 MAbs, 1 directed to a hapten [341] and 2 MAbs binding to epitopes on 2 different protein antigens [568]. In all 3 cases the affinity of the humanized antibodies to the antigen was comparable to that of the original MAb. The fine specificities of the humanized antiprotein MAbs and the tissue distribution of their antigens still have to be tested and compared to the properties of the original rodent MAbs.

If the 'humanized' MAbs turn out to have the same specificities as the donor MAbs, then this technique will provide an opportunity of combining the murine system with its advantages for the generation of specific MAbs against almost any antigen with the attributes of human MAbs and their compatibility for human beings, and will pave the way for a generation of monoclonal antibodies as powerful tools in the diagnosis and therapy of most kinds of human cancer.

This technique especially seems to offer the possibility of constructing fragments of the antibody similar to or smaller than the Fab fragment without impairing the antibody affinity. Such preparations are urgently needed to check whether they further improve immunoscintigraphy and whether they enable radioimmunotherapy of tumors. For tumor immunotherapy, however, antibodies of IgG_1 subtype can be constructed. This subtype possesses strong effector functions [658]. The humanized preparation may probably allow high doses and repetitive treatment without causing allergic problems.

Conclusion

There is no doubt that the invention of the monoclonal antibody technique has brought a new dimension to the possibilities of diagnosis and therapy of tumor diseases. Indeed, a considerable number of antibodies specific for TAA have been generated and developed for immunoscintigraphy as well as for therapy of tumors. We even have preliminary clinical data at our disposal, and the lessons we can learn from these data are by no means completely negative. Thus, specific immunoscintigraphy of tumors is possible. This was shown with gastrointestinal tumors, melanomas and other tumors. Clinical data even indicate that immunoscintigraphy is a technique which provides information relevant to the therapy of the individual patient, which could not be achieved by any other modern technique of tumor diagnosis. However, the clinical experience we have gained has shown us how to improve the antibody for immunoscintigraphy. In addition, the limitations of monoclonal antibodies for any kind of tumor immunotherapy have become apparent (see table LVII).

Results of the dosimetric and pharmacokinetic evaluation of radiolabelled monoclonal antibodies in patients very clearly showed that after i.v. administration 1,000 time less radiolabelled antibody localizes to the tumor site (about 0.015% of the injected material per g tumor) in human patients than in nude mice transplanted with human tumors.

This discovery drew attention to the fact that any increase in the amount of antibody specifically localizing at the tumor site would improve immunoscintigraphy and would improve the possibilities of immunotherapy of tumors (see fig. 18). However, the possibilities of an absolute increase are very restricted: One approach seems to be the generation and development of antibodies specific for an epitope on TAA, which is very selective for tumors and homogeneously expressed by the tumor cells in high amounts ($> 10^6$ epitopes/cell). We have to acknowledge, however, that such TAA might be extremely rare. An additional approach may be the increase in the affinity of the antibody to values higher than Ka 10^8 1/mol.

Table LVII. Valuation of the use of monoclonal antibodies in tumor therapy

Grading	Specific radioimmunotherapy		Specific chemoimmunotherapy		Specific immunotherapy	
	+	−	+	−	+	−
Indications for effectivity						
Systemic treatment	yes		?		yes	
Regional treatment	?		?		yes	
Toxicity						
Systemic treatment		high		high	low	
Regional treatment	moderate		?		low	
Selection of TAA-negative (resistent) cells	no			yes		yes
Amplification system		no		no	yes (anti-idiotype response)	
Practicability		low	moderate		high	
Summary	++ (+)/(+)	− − − −	0 (+)(+)(+)(+)	− − − −	+ + + + + +	−

Induction and selection of such antibodies, however, is extremely difficult, complicated, and also requires a lot of work. The recombinant DNA technique may possibly give us the instrument for site-directed mutagenesis at or around the paratope region of the antibody to improve its affinity. However, this approach is by no means a simple task. Facing the fact that our possibilities of increasing the absolute amount of antibodies localizing at the tumor site are obviously very restricted, it seems essential to increase (at least) its relative amount. One way of doing this might be to construct an antibody preparation which is able to penetrate to the tumor site very quickly (for instance, between 2 and 4 h) but which, in case it is not bound to the tumor (and that is the case for 99.985% of the antibody/g tumor), is then eliminated from the body within a short time.

Another way would be the development of new conjugation procedures, which neither impair the antigen-binding activity of the antibody nor reduce the activity of the cytostatic compound, but by which the localization of the conjugated antibody to normal tissue is drastically reduced.

Experimental as well as clinical data point to the possibility that Fab fragments or preparations smaller than Fab may fulfill the conditions of quick penetration and elimination. However, these data have been obtained by antibodies labelled with iodine radionuclides. It is a well-known fact that iodine is split off the antibody by deiodinases of the RES. In this respect antibody fragments seem to be more sensitive to deiodinases than intact IgG. Since the radiation of iodine radionuclides is measured in dosimetric studies using iodine antibody conjugates, free iodine may simulate a distribution of the conjugate in the body which is not factual. This has already been established for $F(ab')_2$-^{131}I conjugates. Thus experiments with stable conjugates of Fab should be done to confirm the dosimetric studies, gained with iodine-labelled fragments. Proteolytic cleavage of antibodies may damage the antibody binding site and its affinity. Therefore, this procedure does not seem to be optimal for the production of Fab or smaller fragments. An alternative would be the recombinant technique of producing and humanizing antibodies. This technique may enable us to construct fragments even smaller than Fab.

It is obvious that the radiolabelling of these fragments without damaging the binding site with radionuclides suitable for immunoscintigraphy or radiotherapy is yet another critical step, which must still be improved upon.

Altogether, an increase in the absolute and relative amount of antibodies localizing at the tumor site might be possible and should be approached in the future. It seems to us that the preserving production of

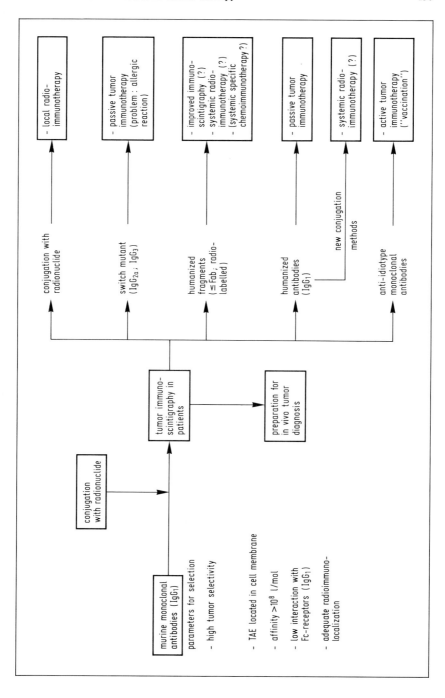

Conclusion

Fab fragments can be achieved via the recombinant DNA technique, with which it is already possible to transplant the murine DNA coding for the paratope (CDR) region into a human DNA coding for the framework and constant regions of immunoglobulins.

This technique additionally enables us to humanize clinically approved murine monoclonal antibodies to get rid of the allergic problems induced by the xenogeneity of the protein. However, a considerable part of the patient's immune response towards murine immunoglobulin is directed at the idiotype.

Up to now it is uncertain which role the stimulation of the idiotype/anti-idiotype network response may play in tumor patients. However, its occurrence simultaneously with delayed tumor regressions indicates that there might be a causal relationship. If so, we would have the key to generate the patient's own antibody response to epitopes on his/her own tumors which before that had been recognized by the mouse to be selective for that tumor.

The generation of the patient's own antibody response by that key would also mean that we have at our disposal an amplification system in the patient and that the quantitative problem is solved if the amount of antibody localizing at the tumor site is sufficient for successful immunotherapy. However, another problem may arise: the isotype of that anti-idiotype and anti-anti-idiotype antibodies may not sufficiently activate the effector function. In that case such antibodies bound to the tumor cell would cover the epitope on the tumor cell and would prevent any other epitope-specific immune response.

It is known that the tumor cell is endowed with a significant potency in qualitatively as well as quantitatively changing its membrane antigens including the TAA. In view of this fact, any passive (application of murine monoclonal antibodies) or active (stimulation of the idiotype/anti-idiotype system) immunotherapy of the tumor with monoclonal antibodies directed against epitopes on TAA brings with it the risk of selecting tumor cells which do not express this specific epitope and are thus resistant to this type of treatment. This problem can be overcome by conjugating a cytotoxic agent to an antibody whose cytotoxic potency is not restricted to the tumor cell the carrier antibody is binding to, but also affects neighboring cells, including TAA-negative tumor cells. Radionuclides might be the cytotoxic

Fig. 18. Network of research on monoclonal antibodies in tumor therapy.

agent of choice. Its use in radioimmunotherapy of tumors, however, is hampered by the fact that the amount of immunoconjugate accumulation at the tumor site in man is extremely low and that about 99.985% of the injected immunoconjugate (in case the patient's tumor mass is about 1 g) damages normal tissues.

As a consequence, future research in tumor therapy using monoclonal antibodies must concentrate on the goal of improving the amount of antibody localizing at the tumor site while at the same time reducing the amount of antibody binding to normal tissues and remaining in the blood pool.

Acknowledgement

The authors are very much indebted to Mrs. S. Lehnert, who typed the manuscript and did all the organizational work with extraordinary skill.

References

1. Abrams, P.G.; Knost, J.A.; Clarke, G.; Wilburn, S.; Oldham, R.K.; Foon, k.A.: Detection of the optimal human cell lines for development of human hybridomas. J. Immunol. *131:* 1201 (1983).
2. Abrams, P.G.; Ochs, J.J.; Giardina, S.L.; Morgan, A.C.; Wilburg, S.B.; Wilt, A.R.; Oldham, R.K.; Foon, K.A.: Production of large quantities of human immunoglobulin in the ascites of athymic mice: Implications for the development of anti-human idiotype monoclonal antibodies. J. Immunol. *132:* 1611 (1984).
3. Adams, D.O.; Hall, T.; Steplewski, Z.; Koprowski, H.: Tumors undergoing rejection induced by monoclonal antibodies of the IgG2a isotype contain increased numbers of macrophages activated for a distinctive form of antibody-dependent cytolysis. Proc. natn. Acad. Sci. USA *81:* 3506 (1984).
4. Adams, D.O.; Lewis, J.G.; Johnson, W.J.: Analysis of interactions between immunomodulators and mononuclear phagocytes: Different modes of tumor cell injury require different forms of macrophage activation. Behring Inst. Mitt. *74:* 132 (1984).
5. Adelstein, S.J.; Kassis, A.I.: Radiobiologic implications of the microscopic distribution of energy from radionuclides. Nucl. Med. Biol. *14:* 165 (1987).
6. Allum, W.H.; MacDonald, F.; Anderson, P.; Fielding, J.W.L.: Localisation of gastrointestinal cancer with a 131 I labelled monoclonal antibody to CEA. Br. J. Cancer *53:* 203 (1986).
7. Alvarez, V.L.; Wen, M.-L.; Lee, C.; Popes, A.D.; Rodwell, J.D.; McKearn, T.J.: Site-specifically modified ^{111}In labelled antibodies give low liver backgrounds and improved radioimmunoscintigraphy. Int. J. Radiat. Appl. Instrum. Part B Nucl. Med. Biol. *13:* 347 (1986).
8. Amzel, L.M.; Poljak, R.J.: Three-dimensional structure of immunoglobulins. Ann. Rev. Biochem. *48:* 961 (1979).
9. Anderson, C.L.; Looney, R.J.: Human leukocyte IgG Fc receptors. Immunology Today *7:* 264 (1986).
10. Anderson-Berg, W.T.; Squire, R.A.; Strand, M.: Specific radioimmunotherapy using ^{90}Y-labelled monoclonal antibody in erythroleukemic mice. Cancer Res. *47:* 1905 (1987).
11. Andres, R.Y.; Schubiger, P.A.: Radiolabelling of antibodies: Methods and limitations. Nucl. Med. *25:* 162 (1986).
12. Arano, Y.; Yokoyama, A.; Magata, Y.; Saji, H.; Horiuchi, K.; Torizuka, K.: Synthesis and evaluation of a new bifunctional chelating agent for Tc-99m-labelling proteins: CE-DTS. Int. J. nucl. Med. Biol. *12:* 425 (1986).
13. Arklie, J.; Taylor-Papadimitriou, J.; Bodmer, W.; Egan, M.; Mellis, R.: Differentiation antigens expressed by epithelial cells in lactating breast are also detectable in breast cancers. Int. J. Cancer *28:* 23 (1981).

14 Armitage, N.C.; Perkins, A.C.; Hardcastle, J.D.; Pimm, M.V.; Baldwin, R.W.: Monoclonal antibody imaging and benign gastrointestinal disease; in Baldwin, Byers, Monoclonal antibodies for cancer detection and therapy, p. 129 (Academic Press, London 1985).

15 Armitage, N.C.; Perkins, A.C.; Pimm, M.V.; Wastie, M.L.; Baldwin, R.W.; Hardcastle, J.D.: Imaging of primary and metastatic colorectal cancer using an ^{111}In- labelled antitumor monoclonal antibody (791T/36). Nucl. Med. Comm. 6: 623 (1985).

16 Arnon, R.: Antibodies and dextran as anti-tumour drug carriers; in Gregoriadis, Trouet, Targeting of drugs p. 31 (Plenum Press, New York 1982).

17 Arnon, R.; Sela, M.: In vitro and in vivo efficacy of conjugates of daunomycin with antitumor antibodies. Immunol. Rev. 62: 5 (1982).

18 Ashorn, R.; Ashorn, P.; Punnonen, R.; Ponyhoenen, L.; Turjanmaa, V.; Koskinen, M.; Helle, M.; Unsitalo, A.; Pystynen, P.; Krohn, K.: The use of radiolabelled monoclonal antibodies to human milk fat globule membrane antigens in antibody-guided tumor imaging, and administration of therapeutic dose of labelled antibody in wide spread ovarian cancer. A preliminary report. Ann. Chir. Gynaecol. 197: suppl. p. 5 (1985).

19 Atcher, R.W.; Gansow, O.A.; Brechbiel, M.B.; Fitzgerald, J.B.: A general method for labelling proteins with trivalent metal-EDTA. Proc. Am. Chem. Soc., Div. of Nucl. Chemistry and Technology, May 1985, Miami Beach, USA; Abstract No. 33.

20 Atkinson, B.F.; Ernst, C.S.; Herlyn, M.; Steplewski, Z.; Sears, H.F.; Koprowski, H.: Gastrointestinal cancer-associated antigen in immunoperoxidase assay. Cancer Res. 42: 4820 (1982).

21 Atkinson, B.; Ernst, C.S.; Ghrist, B.F.D.; Ross, A.H.; Clark, W.H.; Herlyn, M.; Herlyn, D.; Maul, G.; Steplewski, Z.; Koprowski, H.: Monoclonal antibody to a highly glycosylated protein reacts in fixed tissue with melanoma and other tumors. Hybridoma 4: 243 (1985).

22 Badger, C.C.; Krohn, K.A.; Peterson, A.V.; Shulman, H.; Bernstein, I.D.: Experimental radiotherapy of murine lymphoma with 131-I-labelled anti-Thy 1.1 monoclonal antibody. Cancer Res. 45: 1536 (1985).

23 Baldwin, R.W.: Immunity to methylcholanthrene-induced tumors in inbred rats following atrophy and regression of the implanted tumors. Br. J. Cancer 9: 652 (1955).

24 Baldwin, R.W.; Pimm, M.V.: Antitumor monoclonal antibodies for radioimmunodetection of tumors and drug targeting. Cancer Metastasis Rev. 2: 89 (1983).

25 Baldwin, R.W.; Byers, V.S.: Monoclonal antibodies in cancer treatment. Lancet i: 603 (1986).

26 Bale, W.F.; Spar, I.L.; Goodland, R.L.; Woolfe, D.E.: In vivo and in vitro studies of labeled antibodies against rat kidney and walker carcinoma. Proc. Soc. exp. Biol. Med. 89: 564 (1955).

27 Bale, W.F.; Countreras, M.A.; Grady, E.D.: Factors influencing localization of labelled antibodies in tumors. Cancer Res. 40: 2960 (1980).

28 Ball, E.D.; Mills, L.E.; Caughlin, C.T.; Beck, J.R.; Cornwell, G.G.: Autologous bone marrow transplantation in acute myelogenous leukemia: in vitro treatment with myeloid cell-specific monoclonal antibodies. Blood 68: 1311 (1986).

29 Bares, R.; Bockisch, A.; Faß, J.; Oehr, P.; Bull, U.; Biersack, H.J.; Schumpelick, V.: Methodik und Wert der SPECT Technik bei der Radioimmunszintigraphie mit In-111 markierten monoklonalen Antikörperfragmenten am Beispiel des BW MAb 431/31 (anti CEA). Nucl. Med. 26: suppl., p. 43 (1987).

References

30 Barratt, G.M.; Ryman, B.E.; Begent, R.H.J.; Keep, P.A.; Searle, F.; Boden, J.; Bagshawe, K.D.: Improved radioimmunodetection of tumour using liposome-entrapped antibody. Biochim. biophys. Acta 762: 154 (1983).

31 Bartal, A.H.; Feit, C.; Erlandson, R.; Hirshaut, Y.: The presence of viral particles in hybridoma clones secreting monoclonal antibodies. New Engl. J. Med. 306: 1423 (1982).

32 Bartal, A.H.; Feit, C.; Erlandson, R.A.; Hirshaut, Y.: Detection of retroviral particles in hybridomas secreting monoclonal antibodies. Med. Microbiol. Immunol. 174: 325 (1986).

33 Basham, T.Y.; Bourgeade, M.F.; Creasey, A.A.; Merigan, T.C.: Interferon increases HLA synthesis in melanoma cells: interferon-resistant and -sensitive cell lines. Proc. nat. Acad. Sci. USA. 79: 3265 (1982).

34 Bast, R.C.: Feeney, M.; Lazarus, H.; Nadler, L.M.; Colvin, R.B.; Knapp, R.C.: Reactivity of a monoclonal antibody with human ovarian carcinoma. J. clin. Invest. 68: 1331 (1981).

35 Bast, R.C., Jr.; Ritz, J.; Lipton, J.M.; Feeney, M.; Sallan, S.C.; Nathan, D.G.; Schlossman, S.F.: Elimination of leukemic cells from human bone marrow using monoclonal antibody and complement. Cancer Res. 43: 1389 (1983).

36 Bast, R.C., Jr.; De Fabritis, P.; Lipton, J.; Gelber, R.; Maver, C.; Nadler, L.; Sallan, S.; Ritz, J.: Elimination of malignant clonogenic cells from human bone marrow using multiple monoclonal antibodies and complement. Cancer Res. 45: 499 (1985).

37 Baum, R.P.; Maul, F.D.; Senekowitsch, R.; Chatal, J.F.; Saccavini, J.C.; Lorenz, M.; Happ, J.; Berthold, F.; Kriegel, H.; Hör, G.: Radioimmunoscintigraphy and radioimmunotherapy with monoclonal antibodies (19-9/anti-CEA and OC 125); in Höfer, Bergmann, Radioaktive Isotope in Klinik und Forschung. Band 17, Teil 1, p. 443, (Egermann, Wien 1986).

38 Baum, R.P.; Lorenz, M.; Senekowitsch, R.; Albrecht, M.; Hör, G.: Klinische Ergebnisse der Immunszintigraphie und Radioimmuntherapie. Nucl. Med. 26: 68 (1987).

39 Baum, R.P.; Hertel, A.; Madry, N.; Chatenoud, L.; Reynold, J.C.; Saccavini, J.C.; Auerbach, B.; Larson, S.M.; Hör, G.: Human anti-mouse antibody (HAMA) response: Short and longterm follow-up in 76 patients after diagnostic and therapeutic application of radiolabeled monoclonal antibodies. Nucl. Med. (in press).

40 Begent, R.H.J.; Searle, F.; Stanway, G.; Jewkes, R.F.; Jones, B.E.; Vernon, P.; Bagshawe, K.D.: Radioimmunolocalization of tumours by external scintigraphy after administration of 131 I antibody to human chorionic gonadotropin: preliminary communication. J. R. Soc. Med. 73: 624 (1980).

41 Begent, R.H.J.; Keep, P.A.; Green, A.J.; Searle, F.; Bagshawe, K.D.; Jewkes, R.F.; Jones, B.E.; Barratt, G.M.; Ryman, B.E.: Liposomally entrapped second antibody improves tumour imaging with radiolabelled (first) anti-tumour antibody. Lancet ii: 739 (1982).

42 Begent, R.H.J.: Recent advances in tumour imaging: Use of radiolabelled antitumour antibodies. Biochim. biophys. Acta 780: 151 (1985).

43 Beierwaltes, W.H.: Effects of some ^{131}I tagged antibodies in human melanoblastoma: Preliminary report. M. Mich. Med. Bull. 20: 284 (1956).

44 Belles-Isles, M.; Pagé, M.: Anti-oncofoetal proteins for targeting cytotoxic drugs. Int. J. Immunopharmacol. 3: 97 (1981).

45 Belles-Isles, M.; Pagé, M.: In vitro activity of daunomycin-anti-alphafoetoprotein conjugates on mouse hepatoma cells. Br. J. Cancer 41: 841 (1980).

46 (See reference no. 44).
47 Bernal, S. D.; Speak, J. A.: Membrane antigen in small cell carcinoma of the lung defined by monoclonal antibody SM1. Cancer Res. *44:* 265 (1984).
48 Bernard, J. L.; Philip, T.; Zucker, J. M.; Gentet, J.C.: Lutz, P.; Plouvier, E.; Bordigoni, P.; Faurot, M. C.; Raybaud, C.: High dose Vincristine and Melphalan with fractionated total body irradiation and bone marrow transplantation (BMT) as consolidation treatment for an unselected group of 65 stage IV neuroblastoma patients. Proc. Am. Soc. Clin. Oncol. *6:* 220 (1987).
49 Bernhard, M.; Hwang, K.M.; Foon, K.A.; Keenan, A:M.; Kessler, R.M.; Frinck, J.M.; Tallam, D. J.; Hanna, M. G.; Peters, L.; Oldham, R. K.: Localization of ^{111}In and ^{125}I-labeled monoclonal antibody in guinea pigs bearing line 10 hepatocarcinoma tumors. Cancer Res. *43.* 4429 (1983).
50 Bernier, L.G.; Page, M.; Gaudreault, r.C.; Joly, L.P.: A chlorambucil-anti-CEA conjugate cytotoxic for human colon adenocarcinoma cells in vitro. B. J. Cancer *49:* 245 (1984).
51 Berson, Y.: Quantitative aspects of the reaction between insulin-binding antibody. J. clin. Invest. *38:* 1996 (1959).
52 Bhattacharya, M.; Chatterjee, S.K.; Barlow, J.J.; Fuji, H.: Monoclonal antibodies recognizing tumor-associated antigen of human ovarian mucinous cystadenocarcinomas. Cancer Res. *42:* 1650 (1982).
53 Bhattacharya, M.; Chatterjee, S. K.; Gangopadhyay, A.; Barlow, J. J.: Production and characterization of monoclonal antibody to a 60-kD glycoprotein in ovarian carcinoma. Hybridoma *4:* 153 (1985).
54 Biersack, H.J.; Bockisch, A.; Oehr, P.; Knoblich, A.; Hartlapp, J.; Biltz, H.; Jaeger, N.: Bellmann, O.; Vogel, J.; Björklund, B.; Taylor-Papadimitriou, J.; Winkler, C.: Clinical results of immunoscintigraphy in a variety of malignant tumors with special reference to immunohistochemistry. Nuklearmedizin *25:* 167 (1986).
55 Black, C.D.V.; Atcher, R.W.; Barbet, J.; Brechbiel, M.W.; Holton, O.D., III.; Hines, J. J.; Gansow, O. A.; Weinstein, J. N.: Selective ablation of B lymphocytes in vivo by an alpha emitter, Bismuth-212, chelated to a monoclonal antibody (in press).
56 Blaineau, C.; Connan, F.; Arnaud, D.; Andrews, P.; Williams, L.; McIlhinney, J.; Avner, P.: Definition of three species-specific monoclonal antibodies recognizing antigenic structures present on human embryonal carcinoma cells which undergo modulation during in vitro differentiation. Int. J. Cancer *34:* 487 (1984).
57 Blair, A.H.; Ghose, T.I.: Linkage of cytotoxic agents to immunoglobulins. J. immunol. Methods *59:* 129 (1983).
58 Blythman, H.E.; Casellas, P.; Gros, O.; Gros, P.; Jansen, F.K.; Paolucci, F.; Pau, B.; Vidal, H.: Immunotoxins: hybrid molecules of monoclonal antibodies and a toxin subunit specifically kill tumour cells. Nature *290:* 145 (1981).
59 Boeri, D. G.: Europathologie: Recherches cliniques sur la respiration, sur le rire, sur le pleurer et sur le baillement des hémiplégiques. Gaz. heb. méd. Chir. *6:* 73 (1901).
60 Bolton, A. E.; Hunter, W. M.: The labelling of proteins to high specific radioactivities by conjugation to ^{125}I-containing alcylating agent. Biochem. J. *133:* 529 (1973).
61 Bona, C.A.; Kohler, H.: Anti-idiotypic antibodies and internal images; in Venter, Fraser, Lindstrom, Probes for receptor structure and function, vol. 4, pp. 141 (Liss, New York 1984).
62 Borch, R.F.; Bernstein, M.D.; Durst, H.D.: The cyanohydrioborate anion as a selective reducing agent. J. Amer. chem. Soc. *93:* 2897 (1971).

63 Borup-Christensen, P.; Erb, K.; Jensenius, J.C.; Nielson, B.; Svehag, S.E.: Human-human hybridomas for the study of antitumour immune response in patients with colorectal cancer. Int. J. Cancer *37:* 683 (1986).

64 Bosslet, K.; Kurrle, R.; Ax, W.; Sedlacek, H.H.: Monoclonal murine antibodies with specificity for tissue culture lines of human squamous-cell carcinoma of the lung. Cancer Detect. Prevent. *6:* 181 (1983).

65 Bosslet, K.; Stark, M.; Kurrle, R.; Bischof, W.; Sedlacek, H.H.: Microenvironmental influences on the expression of monoclonal antibody defined immunochemically characterized membrane associated antigens on human small cell and adenocarcinomas of the lung. Advances in Cancer Research, Proceedings: Int. Soc. for Oncodevelopmental Biol. and Med., XI Annual Meeting, Stockholm, 1983.

66 Bosslet, K.; Hormel, M.; Schmidt, H.; Müller, K.; Sedlacek, H.H.: Identical expression of cell membrane antigens on a human lung cancer cell line and its drug resistant variant as revealed by a battery of monoclonal anti lung cancer antibodies. Proc. 2nd Symp. of the Section of Exp. Cancer Res. (SEK) of the German Cancer Society, Heidelberg, 1983. J. Cancer Res. clin. Oncol. *105:* 2 (1983).

67 Bosslet, K.; Kern, H.F.; von Bülow, M.; Röher, H.D.; Klöppel, G.; Schorlemmer, H.U.; Kurrle, R.; Sedlacek, H.H.: A human monocyte cell surface antigen, highly expressed on an established pancreatic carcinoma cell line (TU II). Proc. Am. Pancreatic Assoc. and the Nat. Pancreatic Cancer Project. Digestive Diseases and Sciences *28:* 928 (1983).

68 Bosslet, K.; Hilfenhaus, J.: Verfahren zur Inaktivierung hüllhaltiger Viren in mittels einer Zelle in vitro hergestellten Proteinpräparationen (in preparation).

69 Bosslet, K.; Lüben, G.; Schwarz, A.; Hundt, E.; Harthus, H.P.; Seiler, F.R.; Muhrer, C.; Klöppel, G.; Kayser, K.; Sedlacek, H.H.: Immunohistochemical localization and molecular characteristics of three monoclonal antibody-defined epitopes detectable on carcinoembryonic antigen (CEA). Int. J. Cancer *36:* 75 (1985).

70 Bosslet, K.; Steinstraesser, A.; Schwarz, A.; Kuhlmann, L.; Kanzy, E.J.; Sedlacek, H.H.: Spontaneous idiotype loss and isotype switch variants in a hybridoma producing an antibody suited for immunoscintigraphy of pancreatic carcinomas. Cancer Detect. Prevent. (in press).

71 Bosslet, K.; Kern, H.F.; Kanzy, E.J.; Steinstraesser, A.; Schwarz, A.; Lüben, G.; Schorlemmer, H.U.; Sedlacek, H.H.: A monoclonal antibody with binding and inhibiting activity towards human pancreatic carcinoma cells. I. Immunohistological and immunochemical characterization of a murine monoclonal antibody selecting for well differentiated adenocarcinomas of the pancreas. Cancer Immunol. Immunother. *23:* 185 (1986).

72 Bosslet, K.; Döring, N.; Seemann, G.; Schulz, G.; Sedlacek, H.H.: Immunological tailoring of monoclonal antibodies (MAb) suited for immunotherapy of pancreatic carcinoma. Int. J. Cancer (in press).

73 Bosslet, K.; Madry, N.; Kübel, R.; Büchler, M.; Muhrer, K.H.; Klapdor, R.; Schulz, G.: Human anti-idiotypic MAbs against the pancreatic carcinoma specific murine MAb BW 494. Int. J. Cancer (in press).

74 Bosslet, K.; Steinstraesser, A.; Schwarz, A.; Harthus, H.P.; Lüben, G.; Kuhlmann, L.; Sedlacek, H.H.: Quantitative considerations supporting the irrelevance of circulating serum CEA for the immunoscintigraphic visualization of CEA expressing carcinomas. (in preparation).

75 Bostwick, D.G.; Roth, K.A.; Evans, C.J.; Barchas, J.D.; Bensch, K.G.: Gastrin-releasing

peptide, a mammalian analog of Bombesin, is present in human neuroendocrine lung tumors. Am. J. Pathol. *117:* 195 (1984).

76 Boulianne, G.L.; Hozumi, N.; Shulman, M.J.: Production of functional chimaeric mouse/human antibody. Nature *312:* 643 (1984).

77 Bourguet, P.; Kerbrat, P.; Gedouin, D.; Herry, J.Y.; Saccavini, J.C.: Immunoscintigraphy in ovarian cancer. Comparison with CT scan, ultrasonography and laparotomy. Nucl. Med. *25:* A42 (1986).

78 Brabon, A.C.; Williams, J.F.; Cardiff, R.D.: A monoclonal antibody to a human breast tumor protein released in response to estrogen. Cancer Res. *44:* 2704 (1984).

79 Bradwell, A.R.; Vaughan, A.; Fairweather, D.S.; Dykes, P.W.: Improved radio-immunodetection of tumours using a second antibody. Lancet. *i:* 247 (1983).

80 Bradwell, A.R.; Vaughan, A.T.M.; Dykes, P.W.: Limitations in localising and killing tumours using radiolabelled antibodies. Nucl. Med. *25:* 245 (1986).

81 Brechbiel, M.W.; Gansow, O.A.; Atcher, R.W.; Schlom, J.; Esteban, J.; Simpson, D.E.; Calcher, D.: Synthesis of 1- (p-Isothiocyanatobenzyl) derivatives of DTPA and EDTA. Antibody labeling and tumor-imaging studies. Inorg. Chem. *25:* 2772 (1986).

82 Bregni, M.; De Fabritiis, P.; Raso, V.; Greenberger, J.; Lipton, J.; Nadler, L.; Rothstein, L.; Ritz, J.; Bast, R.C., Jr.: Elimination of clonogenic tumor cells from human bone marrow using a combination of monoclonal antibody: ricin A chain conjugates. Cancer Res. *46:* 1208 (1986).

83 Breimer, M.E.: Adaptation of mass spectrometry for the analysis of tumor antigens as applied to blood group glycolipids of a human gastric carcinoma. Cancer Res. *40:* 897 (1980).

84 Bremer, E.G.; Levery, S.B.; Sonnino, S.; Ghidoni, R.; Canevari, S.; Kannagi, R.; Hakomori, S.I.: Characterization of a glycosphingolipid antigen defined by the monoclonal antibody MBrl expressed in normal and neoplastic epithelial cells of human mammary gland. J. biol. Chem. *259:* 14773 (1984).

85 Brockhaus, M.; Magnani, J.L.; Herlyn, M.: Monoclonal antibodies directed against the sugar sequence of lacto-N-fucopentaose III are obtained from mice immunized with human tumors. Arch. Biochem. Biophys. *217:* 647 (1982).

86 Brown, J.P.; Hewick, R.M.; Hellstrom I.; Hellstrom, K.E.; Doolittle, R.F.; Dreyer, W.J.: Human melanoma-associated antigen p97 is structurally and functionally related to transferrin. Nature (London) *296:* 171 (1982).

87 Brown, B.A.; Davis, G.L.; Saltzgaber-Muller, J.; Simon, P.; Ho, M.-K.; Shaw, P.S.; Stone, B.A.; Sands, H.; Moore, G.P.: Tumor-specific genetically engineered murine/human chimeric monoclonal antibody. Cancer Res. *47:* 3577 (1987).

88 Brun del Re, G.Ü.; Stern, A.C.; Baumgartner, C.; Knapp, W.; Hirt, A.; Morell, A.; Bucher, U.; Wagner, H.P.: Purging of bone marrow with the monoclonal anti Calla antibodies VIL-Al, VIB-E3 and VIB-C5 as a prerequisite for autologous bone marrow transplantation. Exp. Hematol. *13:* suppl. 17, p. 54 (1985).

89 Buchegger, F.; Haskell, C.M.; Schreyer, M.; Scazziga, B.R.; Randin, S.; Carrel, S.; Mach, J.-P.: Radiolabeled fragments of monoclonal antibodies against carcinoembryonic antigen for localization of human colon carcinoma grafted into nude mice. J. exp. Med. *158:* 413 (1983).

90 Buchegger, F.; Schreyer, M.; Carrel, S.; Mach, J.-P.: Monoclonal antibodies identify a CEA cross-reacting antigen of 95 kD (NCA-95) distinct in antigenicity and tissue distribution from the previously described NCA of 55 kD. Int. J. Cancer *33:* 643 (1984).

References

91 Buckley, R.G.; Searle, F.: An efficient method for labelling antibodies with ^{111}In. FEBS Letters *166:* 202 (1984).
92 Buckman, R.; McIlhinney, R.A.J.; Shephard, V.; Patel, S.; Coombes, R.C.; Neville, A.M.: Elimination of carcinoma cells from human bone marrow. Lancet *ii:* 1428 (1982).
93 Buinauskas, P.; McCredie, J.A.; Brown, E.R.; Cole, W.H.: Experimental treatment of tumors with antibodies. Archs. Surg., Chicago *79:* 432 (1959).
94 Bumol, T.F.; Chee, D.O.; Reisfeld, R.A.: Immunochemical and biosynthetic analysis of monoclonal antibody-defined melanoma-associated antigen. Hybridoma *1:* 283 (1982).
95 Bumol, T.F.; Reisfeld, R.A.: Unique glycoprotein-proteoglycan complex defined by monoclonal antibody on human melanoma cells. Proc. natn. Acad. Sci. USA *79:* 1245 (1982).
96 Bumol, T.F.: Studies with antibody/drug conjugates. Br. J. Cancer (in press).
97 Buraggi, G.L.; Callegaro, I.; Turrin, A.; Cascinelli, N.; Attili, A.; Emanuelli, H.; Gasparini, M.: Deleide, G.; Plassio, G.; Dovis, M.: Immunoscintigraphy with ^{123}I, ^{99m}Tc and ^{111}In-labelled F(ab')$_2$ fragments of monoclonal antibodies to a human high molecular weight melanoma associated antigen. J. nucl. Med. *28:* 283 (1984).
98 Buraggi, G.L.; Callegaro, L.; Mariani, G.; Turrin, A.; Cascinelli, N.; Attili, A.; Bombardieri, E.; Terno, G.; Plassio, G.; Dovis, M.: Imaging with I-131 labelled monoclonal antibodies to a high-molecular-weight melanoma-associated antigen in patients with melanoma: Efficacy of whole immunoglobulin and its F(ab')$_2$ fragments. Cancer Res. *45:* 3378 (1985).
99 Buraggi, G.L.; Turrin, A.; Cascinelli, N.; Gasparini, M.; Attili, A.; Belli, F.; Seregni, E.: Clinical demands on radioimmunodetection of melanoma: The importance of prospective trials. Nucl. Med. *25:* A42 (1986).
100 Buraggi, G.L.; Callegaro, L.; Turrin, A.; Bombardieri, E.; Gennari, L.; Regalia, E.; Doci, R.; Gasparini, M.; Seregni, E.: Radioimmunodetection of colo-rectal carcinoma with an anti-CEA antibody; in Schmidt, Ell, Britton, Nuklearmedizin in Forschung und Praxis, p. 430 (Schattauer, Stuttgart, New York 1986).
101 Burchell, J.; Gendler, S.; Girling, A.; Taylor-Papadimitriou, J.: Development and characterization of monoclonal antibodies to the core protein of tumor associated mucins. Advances in the Application of Monoclonal Antibodies in Clinical Oncology, May 6-8, Hammersmith Hosp. London (1987).
102 Burnet, M.F.: Cancer – A biological approach. IV. Practical applications. Br. Med. J. *i:* 844 (1957).
103 Burnet, M.F.: Implications of cancer immunity. Aust. N.Z. J. Med. *1:* 71 (1973).
104 Burnett, K.G.; Leung, J.P.; Martinis, J.: Human monoclonal antibodies to defined antigens: Towards clinical applications; in Engleman, Foung, Larrick, Raubitschek, Human hybridomas and monoclonal antibodies p. 113 (Plenum, New York 1985).
105 Burtin, P.; Chavanel, G.; Hirsch-Marie, H.: Characterization of a second normal antigen that cross-reacts with CEA. J. Immunol. *111:* 1926 (1973).
106 Burton, D.R.: Immunoglobulin G: Functional sites. Mol. Immunol. *22:* 161 (1985).
107 Cahan, L.D.; Irie, R.I.; Singh, R.; Cassidenti, A.: Identification of human neuroectodermal tumor antigen (OFA-1-1) as ganglioside G_{D2}. Proc. natn. Acad. Sci. USA *80:* 5392 (1982).
108 Calafat, J.; Janssen, H.; Hekman, A.: Mouse monoclonal antibodies direct phagocytosis of tumor cells by human monocytes. Leuk. Res. *10:* 1347 (1986).

109 Campbell, A. M.; McCormack, M. A.; Ross, C. A.; Leake, R. E.: Immunological analysis of the specificity of the autologous humoral response in breast cancer patients. Br. J. Cancer 53: 1 (1986).
110 Carey, T. E.; Kimmel, K. A.; Schwartz, D. R.; Richter, D.E.; Baker, S.R.; Krause, C.J.: Antibodies to human squamous cell carcinoma. Otolaryngol. Head Neck Surg. 91: 482 (1983).
111 Carlsson, J.; Drevin, H.; Axén, R.: Protein thiolation and reversible protein-protein conjugation: N-succinimidyl 3-(-2-pyridyldithio)propionate, a new heterobifunctional reagent. Biochem. J. 173: 723 (1978).
112 Carrasquillo, J. A.; Krohn, K. A.; Beaumier, P.; McGuffin, R. W.; Brown, J. P.; Hellström, K.E.; Hellström, J.; Larson, M.: Diagnosis of and therapy for solid tumors with radiolabelled antibodies and immune fragments. Cancer Treat. Rep. 68: 317 (1984).
113 Carrasquillo, J. A.; Abrams, P. G.; Schroff, R. W.; Keenan, A. M.; Morgan, A. C.; Foon, K. A.; Reynold, J. C.; Perentesis, P.; Horowitz, M.; Larson, S. M.: Improved imaging of metastatic melanoma with high dose 9.2.27 In-111 monoclonal antibody. J. nucl. Med. 26: P67 (1985).
114 Casellas, P.; Brown, J. P.; Gros, O.; Gros, P.; Hellstrom, I.; Jansen, F. K.; Poncelet, P.; Roncucci, R.; Vidal, H.; Hellstrom, K. E.: Human melanoma cells can be killed in vitro by an immunotoxin specific for melanoma associated antigen p97. Int. J. Cancer 30: 437 (1982).
115 Casellas, P.; Bourrie, B. J. P.; Gros, P.; Jansen, F. K.: Kinetics of cytotoxicity induced by immunotoxins. J. biol. Chem. 259: 9359 (1984).
116 Ceriani, R. L.; Blank, E. W.; Peterson, J. A.: Experimental immunotherapy of human breast carcinomas implanted in nude mice with a mixture of monoclonal antibodies against human milk fat globule components. Cancer Res. 47: 532 (1987).
117 Chanachai, W.; Shani, J.; Wolf, W.; Harwig, J. F.; Nakamura, R. M.: Standardization of CDI-mediated DTPA-coupling to IgG and IgG2a antibodies for ^{113m}In labelling. Int. J. nucl. Med. Biol. 12: 289 (1985).
118 Chandrasekaran, E. V.; Davila, M.; Nixon, D. W.; Goldarb, M.; Mendicino, J.: Isolation and structures of the oligosaccharide units of carcinoembryonic antigen. J. biol. Chem. 258: 7213 (1983).
119 Chatal, J. F.; Saccavini, J. C.; Fumoleau, P.; Douillard, J. Y.; Curtet, C.; Kremer, M.; Le Mevel, B.; Koprowski, H.: Immunoscintigraphy of colon carcinoma. J. nucl. Med. 25: 307 (1984).
120 Chatal, J. F.; Saccavini, J. C.; Fumoleau, P.; Tournemaine, N.; Curtet, C.; Kremer, M.; Chetanneau, A.; Peltier, P.: Prospective SPECT imaging detection of recurrences of gynecological carcinoma using ^{131}I-OC 125 or 19-9 F(ab')$_2$ monoclonal antibodies. Nucl. Med. 25: A42 (1986).
121 Chatal, J. F.; Herry, J. Y.; Lahneche, B.; Lapalus, F.; Lumbroso, J.D.; Pecking, A.; Rougier, P.; Baum, R.P.; Klapdor, R.; Maul, F.D.; Ruibal, A.; Setodin, J.: Immunoscintigraphic localisation of gastrointestinal carcinomas and their recurrences; in Schmidt, Ell, Britton, Nuclear medicine in research and practice, p. 424 (Schattauer, Stuttgart, New York 1986).
122 Chatal, J. F.: Comparative prospective detection of carcinoma recurrences with SPECT imaging using radiolabeled monoclonal antibodies, ultrasonography and computed tomography, in Srivastava Radiolabeled monoclonal antibodies for imaging and therapy. Potential, problems, and prospects (Pergamon Press, New York 1987).

References

123 Chatenoud, L.; Baudrihaye, M.F.; Chkoff, N.; Kreis, H.; Goldstein, G.; Bach, J.F.: Restriction of the human in vivo immune response against the mouse monoclonal antibody OKT-3 J. Immunol. *137:* 830 (1986).

124 Chatenoud, L.: The immune response against therapeutic monoclonal antibodies. Immunology Today *7:* 367 (1986).

125 Cheresh, D.A.; Harper, J.R.; Schulz, G.; Reisfeld, R.A.: Localization of the gangliosides GD_2 and GD_3 in adhesion plaques and on the surface of human melanoma cells. Proc. natn. Acad. Sci. USA *81:* 5767 (1984).

126 Cheresh, D.A.; Varki, A.P.; Varki, N.M.; Stallcup, W.B.; Levine, J.; Reisfeld, R.A.: A monoclonal antibody recognizes an O-acylated sialic acid in a human melanoma-associated ganglioside. J. biol. Chem. *259:* 7453 (1984).

127 Cheung, N.K.; Saarinen, U.M.; Neely, J.E.; Landmeier, B.; Donovan, D.; Coccia, P.F.: Monoclonal antibodies to a glycolipid antigen on human neuroblastoma cells. Cancer Res. *45:* 2642 (1985).

128 Cheung, N.K.; Landmeier, B.; Neely, J.; Nelson, A.D.; Abramowsky, C.; Ellery, S.; Adams, R.B.; Miraldi, F.: Complete tumor ablation with iodine-131 radiolabelled disialoganglioside GD-2 specific monoclonal antibody against human neuroblastoma xenografted in nude mice. J. natn. Cancer Inst. *77:* 739 (1986).

129 Chin, J.; Miller, F.: Identification and localization of human pancreatic tumor-associated antigens by monoclonal antibodies to RWP-1 and RWP-2 cells. Cancer Res. *45:* 1723 (1985).

130 Chu, B.C.F.; Whiteley, J.M.: High molecular weight derivatives of methotrexate as chemotherapeutic agents. Molec. Pharmacol. *13:* 80 (1977).

131 Coakham, H.: Advances in Neuro-Oncology, Proc. 'Advances in the applications of monoclonal antibodies in clinical oncology' p. 41, London (1987).

132 Cobb, L.M.; Humm, J.L.: Radioimmunotherapy of malignancy using antibody targeted radionuclides. Br. J. Cancer *54:* 863 (1986).

133 Colcher, D.; Horan Hand, P.; Nuti, M.; Schlom, J.: A spectrum of monoclonal antibodies reactive with mammary tumor cells. Proc. natn. Acad. Sci. USA *78:* 3199 (1981).

134 Colcher, D.; Zalutsky, M.; Kaplan, W.; Kufe, D.; Austin, F.; Schlom, J.: Radiolocalization of human mammary tumours in athymic mice by a monoclonal antibody. Cancer Res. *43:* 736 (1983).

135 Colcher, D.; Carrasquillo, J.A.; Esteban, J.M.; Sugarbaker, P.; Reynolds, J.C.; Siler, K.; Bryant, G.; Larson, S.M.; Schlom, J.: Radiolabeled monoclonal antibody B72.3 localization in metastatic lesions of colorectal cancer patients. Nucl. med. Biol. *14:* 251 (1987).

136 Colcher, D.; Keenan, A.M.; Larson, S.M.; Schlom, J.: Prolonged binding of a radiolabeled monoclonal antibody (B72.3) used for the in situ radioimmunodetection of human colon carcinoma xenografts. Cancer Res. *44:* 5744 (1984).

137 Cole, S.P.C.; Campling, B.G.; Louwman, I.H.; Kozbor, D.; Roder, J.C.: A strategy for the production of human monoclonal antibodies reactive with lung tumor cell lines. Cancer Res. *44:* 2750 (1984).

138 Cole, W.C.; De Nardo, S.J.; Meares, C.F.; McCall, M.J.; De Nardo, G.L.; Epstein, A.L.; O'Brien, H.A.; Moi, M.K.: Serum stability of ^{67}Cu chelates: Comparison with ^{111}In and ^{57}Co. Int. J. Radiat. Appl. Instrum. Part B Nucl. Med. Biol. *13:* 363 (1986).

139 Corvalan, J.R.F.; Axton, C.A.; Brandon, D.R.; Smith, W.; Woodhouse: Classification of anti-CEA monoclonal antibodies. Protides biol. Fluids *31:* 921 (1984).

140 Cote, R.J.; Houghton, A.N.: The generation of human monoclonal antibodies and their use in the analysis of the humoral immune response to cancer; in Engleman, Foung, Larrick, Raubitschek, Human hybridomas and monoclonal antibodies, p. 189 (Plenum Press, New York 1985).

141 Cote, R.J.; Morrissey, D.M.; Houghton, A.N.; Beattie, Jr.; E.J.; Oettgen, H.F.; Old, L.J.: Generation of human monoclonal antibodies reactive with cellular antigens. Proc. natn. Acad. Sci. USA 80: 2026 (1983).

142 Cote, R.J.; Morrissey, D.M.; Oettgen, H.F.; Old, L.J.: Analysis of human monoclonal antibodies derived from lymphocytes of patients with cancer. Fed. Proc. 43: 2465 (1984).

143 Cote, R.J.; Morrissey, D.M.; Houghton, A.N.; Thomson, T.M.; Daly, M.E.; Oettgen, H.F.; Old, L.J.: Specificity analysis of human monoclonal antibodies reactive with cell surface and intracellular antigens. Proc. natn. Acad. Sci. USA 83: 2959 (1986).

144 Courtenay-Luck, N.S.; Epenetos, A.A.; Moore, R.; Larche, M.; Pectasides, D.; Dhokia, B.; Ritter, M.A.: Development of primary and secondary immune responses to mouse monoclonal antibodies used in the diagnosis and therapy of malignant neoplasms. Cancer Res. 46: 6489 (1986).

145 Courtenay-Luck, N.S.; Epenetos, A.A.; Winearls, C.G.; Ritter, M.A.: Pre-existing human anti-murine immunoglobulin reactivity due to polyclonal rheumatoid factors. Cancer Res. 47: 4520 (1987).

146 Davies, A.G.; Bourne, S.P.; Richardson, R.B.; Czudek, R.; Wallington, T.B.; Kemshead, J.T.; Coakham, H.B.: Pre-existing anti-mouse immunoglobulin in a patient receiving ^{131}I-murine monoclonal antibody for radioimmunolocalisation. Br. J. Cancer 53: 289 (1986).

147 De Fabritis, P.; Bregni, M.; Lipton, J.; Greenberger, J.; Nadler, L.; Rothstein, L.; Korbling, M.; Ritz, J.; Bast, R.C.: Elimination of clonogenic Burkitt's lymphoma cells from human bone marrow using 4-hydroperoxycyclophosphamide in combination with monoclonal antibodies and complement. Blood 65: 1064 (1985).

148 Delaloye, B.; Bischof-Delaloye, A.; Buchegger, F.; von Fliedner, V.; Grob, J.P.; Volant, J.C.; Pettavel, J.; Mach, J.P.: Detection of colorectal carcinoma by emission-computerized tomography after injection of ^{123}I-labeled Fab or F(ab')$_2$ fragments from monoclonal anti-carcinoembryonic antigen antibodies. J. clin. Invest. 77: 301 (1986).

149 Deguchi, T.; Ming Chu, T.; Leong, S.S.; Horozsewicz, J.S.; Lee, C.-L.: Potential therapeutic effect of Adriamycin-monoclonal anti-prostatic acid phosphatase antibody conjugate on human prostate tumor. J. Urol. 137: 353 (1987).

150 De Leij; Poppema, S.; Nulend, J.K.; Ter Haar, J.G.; Schwander, E.; The, T.H.: Immunoperoxidase staining on frozen tissue sections as a first screening assay in the preparation of monoclonal antibodies directed against small cell carcinoma of the lung. Eur. J. Cancer Clin. Oncol. 20: 123 (1984).

151 Denkers, E.Y.; Badger, C.C.; Ledbetter, J.A.; Bernstein, I.D.: Influence of antibody isotype on passive serotherapy of lymphoma. J. Immunol. 135: 2183 (1985).

152 Diamond, B.; Yelton, D.E.; Scharff, M.D.: Monoclonal antibodies: A new technology for producing serologic reagents. New Engl. J. Med. 304: 1344 (1981).

154 Diener, E.; Diner, U.E.; Sinha, A.; Zie, Z.; Vergidis, R.: Specific immunosuppression by immunotoxins containing daunomycin. Science 231: 148 (1986).

155 Dillmann, R.O.; Shawler, D.L.; Dillman, J.B.; Royston, J.: Therapy of chronic lymphocytic leukemia and cutaneous T-cell lymphoma with T101 monoclonal antibody. J. Clin. Oncol. 2: 881 (1984).

156 Dillman, R.O.; Shawler, D.L.; Dillman, J.B.; Clutter, M.; Wormsley, S.B.; Markman, M.; Frisman, D.: Monoclonal antibody therapy of cutaneous T-cell lymphoma (CTCL). Blood 62: suppl. 1, pp. 212 (1983).
157 Dillman, R.O.; Shawler, D.L.; Sobol, R.E.; Collins, H.F.; Beauregard, J.C.; Wormsley, S.B.; Royston, J.: Murine monoclonal antibody therapy in two patients with chronic lymphocytic leukemia. Blood 59: 1036 (1982).
158 Dillmann, R.O.; Beauregard, J.C.; Shawler, D.L.; Sobol, R.E.; Royston, I.: Results of early trials using murine monoclonal antibodies as anti-cancer therapy; in Peeters, Protides of the biological fluids. Proc. of the 30th Colloquium, 1982, p. 353 (Pergamon Press, Oxford 1983).
159 Dillmann, R.O.; Sawler, D.L.; Dillmann, J.B.; Royston, J.; Clutter, M.: Monoclonal antibody therapy of chronic lymphocytic leukemia. Blood 62: suppl. 1, p. 200a (1983).
160 Dillman, R.O.; Royston, I.: Applications of monoclonal antibodies in cancer therapy. Br. med. Bull. 40: 240 (1984).
161 Dillman, R.O.; Beauregard, J.C.; Shawler, D.L.; Halpern, S.E.; Markman, M.; Ryan, K.P.; Baird, S.M.; Clutter, M.: Continuous infusion of T101 monoclonal antibody in chronic lymphocytic leukemia and cutaneous T-cell lymphoma. J. Biol. Resp. Modif. 5: 392 (1986).
162 Dippold, W.G.; Lloyd, K.O.; Li, L.T.C.; Ikeda, H.; Oettgen, F.; Old, L.J.: Cell surface antigens of human malignant melanoma: Definition of six antigenic systems with monoclonal antibodies. Proc. natn. Acad. Sci. USA 77: 6114 (1980).
163 Dippold, W.G.; Knuth, A.; Meyer zum Büschenfelde, K.H.: Inhibition of human melanoma cell growth in vitro by monoclonal anti GD_3-ganglioside antibody. Cancer Res. 44: 806 (1984).
164 Dippold, W.G.; Knuth, K.R.A.; Meyer zum Büschenfelde, K.H.: Inflammatory tumor response to monoclonal antibody infusion. Eur. J. Cancer Clin. Oncol. 21: 907 (1985).
165 Doherty, P.W.; Griffin, T.; Rusckowski, M.; Gionet, M.; Hunter, R.; Hnatowich, D.J.: The potential utility of ^{111}In labeled OC 125 antibody in patients with gynecological tumors. J. nucl. Med. 27: 881 (1986).
166 Dorfmann, N.A.: The optimal technological approach to the development of human hybridomas. J. Biol. Resp. modif. 4: 213 (1985).
167 Douay, L.; Gorin, N.C.; Lopez, M.; Casellas, P.; Liance, M.C.; Jansen, F.K.; Voisin, G.A.; Baillou, C.; Laporte, J.P.; Najman, A.; Duhamel, G.: Evidence for absence of toxicity of T101 immunotoxin on human hematopoietic progenitor cells prior to bone marrow transplantation. Cancer Res. 45: 438 (1985).
168 Douillard, J.Y.; Lehur, P.A.; Vignoud, J.; Blottiere, H.; Maurel, C.; Thedrez, P.; Kremer, M.; Le Mevel, B.: Monoclonal antibodies specific immunotherapy of gastrointestinal tumors. Hybridoma 5: suppl. 1, p. 139 (1986).
169 Drewinko, B.; Yang, L.Y.; Chan, J.; Trujillo, J.M.: New monoclonal antibodies against colon cancer-associated antigens. Cancer Res. 46: 5137 (1986).
170 Drebin, J.A.; Link, V.C.; Weinberg, R.A.; Greene, M.I.: Inhibition of tumor growth by a monoclonal antibody reactive with an oncogene-encoded tumor antigen. Proc. natn. Acad. Sci. USA 83: 9129 (1986).
171 Dykes, P.W.; Hine, K.R.; Bradwell, A.R.; Blackburn, J.C.; Reeder, T.A.; Drolc, Z.; Booth, S.N.: Localisation of tumour deposits by external scanning after injection of radiolabelled anti-carcinoembryonic antigen. Br. Med. J. 1: 220 (1980).
172 Edmond, S.K.; Grady, L.T.; Outschoorn, A.S.; Rhodes, C.T.: Monoclonal antibodies as drugs or devices: Practical and regulatory aspects. Drug Dev. ind. Pharm. 12: 107 (1986).

173 Edwards, P. A. W.: Heterogenous expression of cell-surface antigens in normal epithelia and their tumours, revealed by monoclonal antibodies. Br. J. Cancer *51:* 149 (1985).
174 Egan, M. L.; Coligan, J. E.; Pritchard, D. G.; Schnute, W. C., Jr.; Tood, C. W.: Physical characterization and structural studies of the carcinoembryonic antigen. Cancer Res. *36:* 3482 (1976).
175 Egan, M. L.; Henson, D. E.: Monoclonal antibodies and breast cancer. J. natn. Cancer Inst. *68:* 338 (1982).
176 Ehrlich, P.: A general review of the recent work in immunity; in Collected papers of Paul Ehrlich, vol. 2: Immunology and cancer research, p. 442 (Pergamon Press, London 1956).
177 Eiklid, K.; Olsnes, S.; Pihl, A.: Entry of lethal doses of abrin, ricin and modeccin into the cytosol of HeLa cells. Exp. Cell Res. *126:* 321 (1980).
178 Embleton, M. J.; Gunn, B.; Byers, V. S.; Baldwin, R. W.: Antitumour reactions of monoclonal antibody against a human osteogenic sarcoma cell line. Br. J. Cancer *43:* 582 (1981).
179 Embleton, M. J.; Rowland, G. F.; Simmonds, R. G.; Jacobs, E.; Marsden, C. H.; Baldwin, R. W.: Selective cytotoxicity against human tumour cells by a vindesine-monoclonal antibody conjugate. Br. J. Cancer *47:* 43 (1983).
180 Embleton, M. J.; Habib, N. A.; Garnett, M. C.; Wood, C.: Unsuitability of monoclonal antibodies to oncogene proteins for antitumour drug-targeting. Int. J. Cancer *38:* 821 (1986).
182 Epenetos, A. A.; Britton, K. E.; Mather, S.; Shepherd, J.; Granowska, M.; Taylor-Papadimitriou, J.; Nimmon, C. C.; Durbin, H.; Hawkins, L. R.; Malpas, J. S.; Bodmer, W. F.: Targeting of iodine-123-labelled tumour-associated monoclonal antibodies to ovarian, breast and gastrointestinal tumours. Lancet *ii:* 999 (1982).
183 Epenetos, A. A.; Snook, D.; Hooker, G.; Begent, R.; Durbin, H.; Oliver, R. T.; Bodmer, W. F.; Lavender, J. P.: [111]Indium labelled monoclonal antibody to placental alkaline phosphatase in the detection of neoplasms of testis, ovary, and cervix. Lancet *ii:* 350 (1985).
184 Epenetos, A. A.; Snook, D.; Durbin, H.; Johnson, P. M.; Taylor-Papadimitriou, J.: Limitations of radiolabeled monoclonal antibodies for localization of human neoplasms. Cancer Res. *46:* 3183 (1986).
185 Epenetos, A. A.: Antibody guided diagnosis and therapy. Proc. 'Advances in the applications of monoclonal antibodies in clinical oncology', p. 42, London (1987).
186 Erlanger, B. F.; Beiser, S. M.; Borek, F.; Edel, F.; Lieberman, S.: The preparation of steroid-protein conjugates to elicit antihormonal antibodies. Methods Immunol. Immunochem. *1:* 144 (1967).
187 Erlanger, B. F.: Anti-idiotypic antibodies: what do they recognize? Immunology Today *6:* 10 (1985).
188 Ernst, C. S.; Shen, J.-W.; Litwin, S.; Herlyn, M.; Koprowski, H.; Sears, H. F.: Multiparameter evaluation of the expression in situ of normal and tumor-associated antigens in human colorectal carcinoma. J. natn. Cancer Inst. *77:* 387 (1986).
189 Eto, Y., Shinoda, S.: Gangliosides and neutral glycosphingolipids in human brain tumors: Specificity and their significance. New vistas in glycolipid research. Adv. Exp. Med. Biol. Monogr. *152:* 279 (1982).
190 Ettinger, D. S.; Order, S. E.; Wharam, M. D.; Parker, M.K.; Klein, J.L.; Leichner, P.K.: Phase I–II study of isotopic immunoglobulin therapy for primary liver cancer. Cancer Treat. Rep. *66:* 289 (1982).

191 Fairweather, D.S.; Bradwell, A.R.; Dykes, P.W.; Vaugham, A.T.; Watson-James, S.F.; Chandler, S.: Improved tumour localisation using indium-111 labelled antibodies. Br. Med. J. *286:* 168 (1983).

192 Fairweather, D.S.; Bradwell, A.R.; Dykes, P.W.: Nuclear imaging techniques with radiolabelled antibodies. J. Pathol. *141:* 363 (1983).

193 Fargion, S.; Carney, D.; Mulshine, J.; Rosen, S.; Bunn, P.; Jewett, P.; Cuttitta, F.; Gazdar, A.; Minna, J.: Heterogeneity of cell surface antigen expression of human small cell lung cancer detected by monoclonal antibodies. Cancer Res. *46:* 2633 (1986).

194 Farrands, P.A.; Perkins, A.C.; Pimm, M.V.; Hardy, J.D.; Embleton, M.J.; Baldwin, R.W.; Hardcastle, J.D.: Radioimmunodetection of human colorectal cancers by an antitumour monoclonal antibody. Lancet *ii:* 397 (1982).

195 Farrands, P.A.; Perkins, A.; Sully, L.; Hopkins, J.S.; Pimm, M.V.; Baldwin, R.W.; Hardcastle, J.D.: Localisation of human osteosarcoma by antitumour monoclonal antibody. J. Bone Jt Surg. (Br.) *65:* 638 (1983).

196 Points to consider in the manufacture and testing of monclonal antibody products for human use. (Office of Biologics Research and Review, FDA, June, 1987).

197 Feizi, T.; Childs, R.A.: Carbohydrates as antigenic determinants of glycoproteins. Biochem. J. *245:* 1 (1987).

198 Fiebig, H.H.; Schuchhardt, C.; Henss, H.; Fiedler, L.; Lohr, G.W.: Comparison of tumor response in nude mice and in the patients. Behring Inst. Mitt. *74:* 343 (1984).

199 Fitzgerald, D.; Trowbridge, I.; Pastan, I.; Willingham, M.: Enhancement of toxicity of anti-transferrin receptor antibody – Pseudomonas exotoxin conjugates by adenovirus. Proc. natn. Acad. Sci. USA *80:* 4134 (1983).

200 Fitzgerald, D.J.; Bjorn, M.J.; Ferris, R.J.; Winkelhake, J.L.; Frankel, A.E.; Hamilton, T.C.; Ozols, R.F.; Willingham, M.C.; Pastan, I.: Antitumor activity of an immunotoxin in a nude mouse model of human ovarian cancer. Cancer Res. *47:* 1407 (1987).

201 Foley, E.J.: Antigenic properties of methylcholanthrene-induced tumors in mice of the strain of origin. Cancer Res. *13:* 835 (1953).

202 Foon, K.A.; Bunn, P.A.; Schroff, R.W.; Mayer, D.; Hsu, S.-M.; Sherwin, S.A.; Oldham, R.K.: Monoclonal antibody serotherapy of chronic lymphocytic leukemia and cutaneous T cell lymphoma: preliminary observations; in Boss, Langmann, Trowbridge, Dulbecco, Monoclonal antibodies and cancer, p. 39 Academic Press, New York 1984).

293 Ford, C.H.J.; Newman, C.E.; Johnson, J.R.; Woodhouse, C.S.; Reeder, T.A.; Rowland, G.F.; Simmonds, R.G.: Localisation and toxicity study of a vindesine-anti-CEA conjugate in patients with advanced cancer. Br. J. Cancer *47:* 35 (1983).

204 Forstrom, J.W.; Nelson, K.A.; Nepom, G.T.; Hellstrom, I.; Hellstrom, K.E.: Immunization to a syngeneic sarcoma by a monoclonal auto-antiidiotypic antibody. Nature *303:* 627 (1983).

205 Foulds, L.: The experimental study of tumor progression: a review. Cancer Res. *14:* 327 (1954).

206 Foung, S.K.H.; Perkins, S.; Arvin, A.; Lifson, J.; Mohagheghpour, N.; Fishwild, D.; Grumet, F.C.; Engleman, E.G.: Production of human monoclonal antibodies using a human mouse fusion partner; in Engleman, Foung, Larrick, Raubitschek, Human hybridomas and monoclonal antibodies, p. 135 (Plenum Press, New York 1985).

207 Frade, R.; Barel, M.; Ehlin-Henriksson, B.; Klein, G.: gp140, the C3d receptor of human B lymphocytes, is also the Epstein-Barr virus receptor. Proc. natn. Acad. Sci. USA *82:* 1490 (1985).

208 Fraker, P. J.; Speck, J. C.: Protein and cell membrane iodinations with a sparingly soluble chloroamide. Biochem. biophys. Res. Commun. *80:* 849 (1978).
209 Frankel, A. E.; Ring, D. B.; Tringale, F.; Hsieh-Ma, S. T.: Tissue distribution of breast cancer-associated antigens defined by monoclonal antibodies. J. Biol. Resp. Modif. *4:* 273 (1985).
210 Frankel, A. E.: Antibody-toxin hybrids: A clinical review of their use. J. Biol. Resp. Modif. *4:* 437 (1985).
211 Fridrich, R.; Andres, R.; Stähli, C.; Zenklusen, H. R.: First experiments on radioimmunodetection with b-12-monoclonal antibody fragments against breast cancer antigen. Nucl. Med. *25:* 225 (1986).
212 Friedman, E.; Thor, A.; Horan Hand, P.; Schlom, J.: Surface expression of tumor-associated antigens in primary cultured human colonic epithelial cells from carcinomas, benign tumors, and normal tissues. Cancer Res. *45:* 5648 (1985).
213 Fritzberg, A. R.: Advances in Tc-99m-labeling of Antibodies. Nucl.-Med. *26:* 7 (1987).
214 Frödin, J. E.; Biberfeld, P.; Christensson, B.; Philstedt, P.; Sundelius, S.; Sylven, M.; Wahren, B.; Koprowski, H.; Mellstedt, H.: Treatment of patients with metastasizing colorectal carcinoma with mouse monoclonal antibodies (Moab 17–1A): A progress report. Hybridoma *5:* suppl. 1, p. 151 (1986).
215 Fukumoto, T.; Brandon, M. R.: The site of IgG2a catabolism in the rat. Mol. Immunol. *18:* 741 (1981).
216 Fukushi, Y.; Hakomori, S.; Nudelman, E.; Cochran, N.: Novel fucolipids accumulating in human adenocarcinoma. II. Selective isolation of hybridoma antibodies that differentially recognize mono-, di-, and trifucosylated type 2 chain. J. biol. Chem. *259:* 4681 (1984).
217 Fukushi, Y.; Nudelman, E.; Levery, S. B.; Rauvala, H.; Hakomori, S. A.: Hybridoma antibody (FH6) defining a human cancer-associated difucoganglioside ($VI^3NeuAcV^3III^3Fuc_2nLc_6$): novel fucolipids accumulating in human cancer. III. J. biol. Chem. *259:* 10511 (1984).
218 Fukushi, Y.; Kannagi, R.; Hakomori, S.-I.; Shepard, T.; Kulander, B. G.; Singer, J. W.: Location and distribution of difucoganglioside ($VI^3III^3Fuc_2nLc_6$) in normal and tumor tissues defined by its monoclonal antibody FH6. Cancer Res. *45:* 3711 (1985).
219 Fukushima, J.; Hirota, M.; Terasaki, P. I.; Wakisaka, A.; Togashi, H.; Chia, D.; Suyama, N.; Fukushi, Y.; Nudelman, E.; Hakomori, S.: Characterization of a new tumor-associated antigen: Sialosylated Lewis. Cancer Res. *44:* 5279 (1984).
220 Gaffar, S. A.; Pant, K. D.; Shochat, D.; Bennett, S. J.; Goldenberg, D. M.: Experimental studies of tumor radioimmunodetection using antibody mixtures against carcinoembryonic antigen (CEA) and colon-specific antigen-p (CSAp). Int. J. Cancer *27:* 101 (1981).
221 Gallego, J.; Price, M. R.; Baldwin, R. W.: Preparation of four daunomycin-monoclonal antibody 191T/36 conjugates with antitumor activity. Int. J. Cancer *33:* 737 (1984).
222 (See reference no. 221).
223a Gallego, J.; Price, M. R.: Monoclonal antibody-drug conjugates: A new approach for cancer therapy. Drugs of today *21:* 511 (1985).
223b Gansow, O. A.; Kauser, A. R.: Diazo coupling of some lanthanide benzo-cryptates to proteins. Inorganica Chimica Acta *91:* 213 (1984).
224 Garnett, M. C.; Embleton, M. J.; Jacobs, E.; Baldwin, R. W.: Preparation and properties of a drug-carrier-antibody conjugate showing selective antibody-directed cytotoxicity in vitro. Int. J. Cancer *31:* 661 (1983).

225 Ghose, T.; Norvell, S.T.; Guclu, A.; Cameron, D.; Bodurtha, D.; MacDonald, A.S.: Immunochemotherapy of cancer with chlorambucil-carrying antibody. Br. Med. J. *3:* 495 (1972).

226 Ghose, T.; Guclu, A.: Cure of a mouse lymphoma with radioiodinated antibody. Eur. J. Cancer *10:* 787 (1974).

227 Ghose, T.; Nigam, S.: Antibody as carrier of chlorambucil. Cancer *29:* 1398 (1972).

228 Ghose, T.; Norvell, S.T.; Guclu, A.; Bodurtha, A.; Tai, J.; MacDonald, A.S.: Immunochemotherapy of malignant melanoma with chlorambucil-bound antimelanoma globulins: preliminary results in patients with disseminated disease. J. natn. Cancer Inst. *158:* 845 (1977).

229 Ghose, T.; Blair, A.H.: Antibody-linked cytotoxic agents in the treatment of cancer. Current status and future prospects. J. natn. Cancer Inst. 61: 657 (1978).

230 Ghose, T.; Norvell, S.T.; Aquino, J.; Belitzky, P.; Tai, J.; Guclu, A.; Blair, A.H.: Localization of ^{131}I-labeled antibodies in human renal cell carcinomas and in a mouse hepatoma and correlation with tumor detection by photoscanning. Cancer Res. *40:* 3018 (1980).

231 Ghose, T.; Blair, A.H.; Uadia, P.; Kulkarni, P.N.; Goundalkar, A.; Mezei, M.; Ferrone, S.: Antibodies as carriers of cancer chemotherapeutic agents. Ann. N.Y. Acad. Sci. *446:* 213 (1985).

232 Ghose, T.; Ramakrishnan, S.; Kulkarni, P.; Blair, A.H.; Vaughan, K.; Nolido, H.; Norvell, S.T.; Belitsky, P.: Use of antibodies against tumor-associated antigens for cancer diagnosis and treatment. Transplant. Proc. *13:* 1970 (1981).

233 Ghose, T.; Ferrone, S.; Imai, K.; Norvell, S.T., Jr.; Luner, S.J.; Martin, R.H.; Blair, A.H.: Imaging of human melanoma xenografts in nude mice with a radiolabelled monoclonal antibody. J. natn. Cancer Inst. *69:* 823 (1982).

234 Ghose, T.; Guclu, A.; Rama, R.R.; Blair, A.H.: Inhibition of a mouse hepatoma by the alkylating agent trenimon linked to immunoglobulins. Cancer Immunol. Immunother. *13.* 185 (1982).

235 Ghose, T.; Kulkarni, N.; Ferrone, S.; Giacomini, P.; Norvell, S.T.; Kulkarni, P.; Blair, A.H.: Imaging human tumors in nude mice; in Burchiel, Rhodes, Radioimmunoimaging and radioimmunotherapy, p. 255 (Elsevier, Amsterdam 1983).

236 Ghose, T.; Blair, A.H.; Kulkarni, P.: Preparation of antibody-linked cytotoxic agents. Methods Enzym. *93:* 280 (1983).

237 Ghose, T.; Blair, A.H.; Vaugham, K.; Kulkarni, P.: Antibody-directed drug targeting in cancer; in Goldberg, Targeted drugs, p. 1 (John Wiley & Sons, New York 1983).

238 Ghose, T.; Blair, A.H.: The design of cytotoxic-agent-antibody conjugates. CRC crit. Rev. Ther. Drug Carrier Systems *3:* 263 (1987).

239 (See reference no. 238).

240 Giacomini, P.; Aguzzi, A.; Pestka, S.; Fisher, P.B.; Ferrone, S.: Modulation by recombinant DNA leukocyte (alpha) and fibroblast (beta) interferons of the expression and shedding of HLA- and tumor-associated antigens by human melanoma cells. J. Immunol. *133:* 1649 (1984).

241 Gilliland, D.G.; Collier, R.J.: A model system involving anticoncanavalin A for antibody targeting of diphtheria toxin fragment Al. Cancer Res.: *40:* 3564 (1980).

242 Gilliland, D.G.; Steplewski, Z.; Collier, R.J.; Mitchell, K.; Chang, T.; Koprowski, H.: Antibody directed cytotoxic agents: Use of monoclonal antibodies to direct the action of toxin A chains to colorectal carcinoma cells. Proc. natn. Acad. Sci. USA *77:* 4539 (1980).

243 Gold, P.; Freedman, S.O.: Demonstration of tumor-specific antigens in human colonic

carcinomata by immunological tolerance and absorption techniques. J. exp. Med. *121:* 439 (1965).

244 Goldenberg, D.M.; Preston, D.F.; Primus, F.J.; Hansen, H.J.: Photoscan localization of GW-39 tumors in hamsters using radiolabeled anticarcinoembryonic antigen immunoglobulin G. Cancer Res. *34:* 1 (1974).

245 Goldenberg, D.M.; DeLand, F.; Kim, E.; Bennett, S.; Primus, F.J.; Van Nagell, J.R.; Estes, N.; De Simone, P.; Rayburn, P.: Use of radiolabeled antibodies to carcinoembryonic antigen for the detection and localization of diverse cancers by external photoscanning. New Engl. J. Med. *298:* 1384 (1978).

246 Goldenberg, D.M.; Gaffar, S.A.; Bennet, S.J.; Beach, J.L.: Experimental radioimmunotherapy of a xenografted human colonic tumor (GW-38) producing carcinoembryonic antigen. Cancer Res. *41:* 4354 (1981).

247 Gallick, G.E.; Kurzrock, R.; Kloetzer, W.S.; Arlinghaus, R.B.; Gutterman, J.N.: Expression of $p21^{ras}$ in fresh primary and metastatic human colorectal tumors. Proc. natn. Acad. Sci. USA *82:* 1795 (1985).

248 Goodman, G.E.; Beaumier, P.; Hellström, H.; Fernyhough, B.; Hellström, K.-E.: Pilot trial of murine monoclonal antibodies in patients with advanced melanoma. J. clin. Oncol. *3:* 340 (1985).

249 Gorin, N.C.; Douay, L.; Laporte, J.P.; Lopez, M.; Zittoun, R.; Rio, B.; David, R.; Stachowiak, J.; Jansen, J.; Cazellas, P.; Poncelet, P.; Liance, M.C.; Coisin, G.A.; Salmon, C.; Le Blanc, G.; Deloux, J.; Najman, A.; Duhamel, G.: Autologous bone marrow transplantation with marrow decontaminated by immunotoxin T 101 in the treatment of leukemia and lymphoma: first clinical observation. Cancer Treat. Rep. *69*: 953 (1985).

250 Granowska, M.; Shepherd, M.; Britton, K.E.; Ward, B.; Mather, S.; Taylor-Papadimitriou, J.; Epenetos, A.A.; Carroll, M.J.; Nimmon, C.C.; Hawkins, L.A.: Ovarian cancer: Diagnosis using ^{123}I monoclonal antibody in comparison with surgical findings. Nucl. Med. Commun. *5:* 485 (1984).

251 Granowska, M; Britton, K.E.; Shepherd, J.H.; Nimmon, C.C.; Mather, S.; Ward, B.; Osborne, R.J.; Slevin, M.L.: A prospective study of ^{123}I labeled monoclonal antibody imaging in ovarian cancer. J. clin. Oncol. *4:* 730 (1986).

252 Greiner, J.W.; Horan Hand, P.; Noguchi, P.; Fisher, P.; Pestka, S.; Schlom, J.: Enhanced expression of surface tumor associated antigens on human breast and colon tumor cells after recombinant leukocyte α-interferon treatment. Cancer Res. *44:* 3208 (1984).

253 Greiner, J.W.; Tobi, M.; Fisher, P.B.; Langer, J.A.; Pestka, S.: Differential responsiveness to cloned mammary carcinoma cell populations to the human recombinant leukocyte interferon enhancement of tumor antigen expression. Int. J. Cancer *36:* 159 (1985).

254 Greiner, J.W.; Guadagni, F.; Noguchi, P.; Pestka, S.; Colcher, D.; Fisher, P.B.; Schlom, J.: Recombinant interferon enhances monoclonal antibody-targeting of carcinoma lesions in vivo. Science *235:* 895 (1987).

255 Griffin, T.W.; Richardson, C.; Houston, L.L.; LePage, D.; Bogden, A.; Raso, V.: Antitumor activity of intraperitoneal immunotoxins in a nude mouse model of human malignant mesothelioma. Cancer Res. *47:* 4266 (1987).

256 Griswold, W.R.: A quantitative relationship between antibody affinity and antibody avidity. Immunol. Invest. *16:* 97 (1987).

257 Gross, L.: Intradermal immunization of C3H mice against a sarcoma that originated in an animal of the same line. Cancer Res. *3:* 326 (1943).

258 Guclu, A.; Tai, J.; Ghose, T.: Endocytosis of chlorambucil-bound anti-tumor globulin following 'capping' in EL4 lymphoma cells. Immunol. Commun. *4*: 299 (1975).
259 Häkkinen, I.: A-like blood group antigen in gastric cancer cells of patients in blood groups 0 and B. J. natn. Cancer Inst. *44*: 1183 (1970).
260 Hagan, P.L.; Halpern, S.E.; Dillman, R.O.; Shawler, D.L.; Johnson, D.E.; Chen, A.; Krishnan, L.; Frincke, J.; Bartholomew, R.M.; David, G.S.; Carlo, D.: Tumor size: effect on monoclonal antibody uptake in tumor models. J. nucl. Med. *27*: 422 (1986).
261a Hakomori, S.; Wang, S.-H.; Young, W.W., Jr.: Isoantigenic expression of Forssman glycolipid in human gastric and colonic mucosa: its possible identity with 'A-like antigen' in human cancer. Proc. natn. Acad. Sci. USA *74*: 3023 (1977).
261b Hakomori, S.-I.: Glycosphingolipids in cellular interaction, differentiation, and oncogenesis. Ann. Rev. Biochem. *50*: 733 (1981).
262 Hakomori, S.; Nudelmann, E.; Kannagi, R.; Levery, S.: The common structure in fucosyllactosaminolipids accumulating in human adenocarcinomas, and its possible absence in normal tissue. Biochem. biophys. Res. Commun. *109*: 36 (1982).
263 Hakomori, S.; Kannagi, R.: Glycosphingolipids as tumor-associated and differentiation markers. J. natn. Cancer Inst. *71*: 231 (1983).
264 Hakomori, S.; Nudelman, E.; Levery, S.B.; Kannagi, R.: Novel fucolipids accumulating in human adenocarcinoma. I. Glycolipids with di- or trifucosylated type 2 chain. J. biol. Chem. *259*: 4672 (1984).
265 Hakomori, S.: Tumor associated carbohydrates antigens. Ann. Rev. Immunol. *2*: 103 (1984).
266 Hakomori, S.I.: Aberrant glycosylation in cancer cell membranes as focused on glycolipids: Overview and perspectives. Cancer Res. *45*: 2405 (1985).
267 Halpern, S.E.; Dillmann, R.O.; Witztum, K.F.; Shega, J.F.; Hagan, P.L.; Burrows, W.M.; Dillman, J.B.; Clutter, M.L.; Sobol, R.E.: Radioimmunodetection of melanoma utilizing In-111 96.5 monoclonal antibody: a preliminary report. Radiology *155*: 493 (1985).
268 Hamblin, T.J.; Catton, A.R.; Glennie, M.J.; MacKenzie, M.R.; Stevenson, F.K.; Watts, H.F.; Stevenson, G.T.: Initial experience in treating human lymphoma with a chimeric univalent derivative of monoclonal anti-idiotype antibody. Blood *69*: 790 (1986).
269 Hamblin, T.J.; Abdul-Ahad, A.K.; Gordon, J.; Stevenson, F.K.; Stevens, G.T.: Preliminary experience in treating lymphocytic leukemia with antibody to immunoglobulin idiotypes on the cell surface. Br. J. Cancer *42*: 495 (1980).
270 Hammarström, S.; Engvall, E.; Johansson, B.G.; Svensson, S.; Sundblad, G.; Goldstein, I.J.: Nature of the tumor-associated determinant(s) of carcinoembryonic antigen. Proc. natn. Acad. Sci. USA *72*: 1528 (1975).
271 Hammarström, S.: Chemistry and immunology of CEA, CA 19-9 and CA-50; in Holgren, Tumor marker antigens, p. 32 (Studentlitteratur, Lund, Sweden 1985).
272 Harkonen, S.; Stoudemire, J.; Mischak, R.; Spitler, L.E.; Lopez, H.; Scannon, P.: Toxicity and immunogenicity of monoclonal antimelanoma antibody-ricin A chain immunotoxin in rats. Cancer Res. *47*: 1377 (1987).
273 Harwood, P.J.; Pedley, R.B.; Boden, J.; Rawlings, G.; Pentycross, C.R.; Rogers, G.T.; Bagshawe, K.D.: Prolonged localisation of a monoclonal antibody against CEA in a human colon tumor xenograft. Br. J. Cancer *52*: 797 (1985).
274 Hashimoto, N.; Takatsu, K.; Masuho, Y.; Kishida, K.; Hara, T.; Hamaoka, T.: Selective elimination of a B cell subset having acceptor site(s) for T cell replacing factor (TRF)

with biotinylated antibody to the acceptor site(s) and avidin-ricin A-chain conjugate. J. Immunol. *132*: 129 (1984).

275 Haskell, L.M.; Buchegger, F.; Schreyer, M.; Carrel, S.; Mach, J.-P.: Monoclonal antibodies to carcinoembryonic antigen: Ionic strength as a factor in the selection of antibodies for immunoscintigraphy. Cancer Res. *43*: 3857 (1983).

276 Haspel, M.V.; McCabe, R.P.; Pomato, N.; Hoover, H.C.; Hanna, M.G.: Human monoclonal antibodies: Generation of tumor-cell reactive monoclonal antibodies using peripheral blood lymphocytes from actively immunised patients with colorectal carcinoma; in Reisfeld, Sell, Monoclonal antibodies and cancer therapy, p. 505 (Liss, New York 1985).

277 Hattori, H.; Uemura, K.; Taketomi, T.: Glycolipids of gastric cancer. The presence of blood group A-active glycolipids in cancer tissues from blood group 0 patients. Biochim. biophys. Acta *666*: 351 (1981).

278 Hatzubai, A.; Maloney, D.G.; Levy, R.: The use of a monoclonal anti-idiotype antibody to study the biology of a human B cell lymphoma. J. Immunol. *126*: 2397 (1981).

279 Havemann, K.; Holle, R.; Jaques, G.; Gropp, C.; Victor, N.; Drings, P.; Manke, H.G.; Hans, K.; Schroeder, M.; Heim, M.: Tumormarker beim kleinzelligen Bronchialkarzinom. Ergebnisse einer prospektiven multizentrischen Studie; in Tumormarker: Aktuelle Aspekte und klinische Relevanz, Wüst, p. 84. (Steinkopff, Darmstadt 1985).

280 Hawkes, R.; Niday, E.; Gordon, J.: A dot-immunobinding assay for monoclonal and other antibodies. Anal. Biochem. *119*: 142 (1982).

281 Hayes, C.E.; Goldstein, I.J.: Radioiodination of sulfhydryl-sensitive proteins. Anal. Biochem. *67*: 580 (1975).

282 Hedin, A.; Hammarström, S.; Svenberg, T.; Sundblad, G.: Low molecular weight CEA, isolation and characterization. Scand. J. Immunol. *8*: suppl. 8, p. 423 (1978).

283 Hedin, A.; Wahren, B.; Hammaström, S.: Tumor localization of CEA-containing human tumors in nude mice by means of monoclonal anti-CEA antibodies. Int. J. Cancer *30*: 547 (1982).

284 Hedin, A.; Hammarström, S.; Larsson, A.: Specificities and binding properties of eight monoclonal antibodies against carcinoembryonic antigen. Mol. Immunol. *19*: 1641 (1982).

285 Hedin, A.; Carlsson, L.; Berglund, A.; Hammarström, S.: A monoclonal antibody-enzyme immunoassay for serum carcinoembryonic antigen with increased specificity for carcinomas. Proc. natn. Acad. Sci. USA *80*: 3470 (1983).

286 Hellström, I.; Brown, J.P.; Klitzman, J.M.; Hellström, K.E.: A highly sensitive, and reproducible microcytotoxicity assay for demonstrating cytotoxic antibodies to cell surface antigens. J. Immunol. Methods *22*: 369 (1987).

287 Hellström, J.; Brankovan, V.; Hellström, K.E.: Strong antitumor activities of IgG3 antibodies to a human melanoma associated ganglioside. Proc. natn. Acad. Sci. USA *82*: 1499 (1985).

288 Hendler, F.J.; Ozanne, B.W.: Human squamous cell lung cancer express increased epidermal growth factor receptors. J. clin. Invest. *74*: 647 (1984).

289 Heppner, G.H.: Tumor heterogeneity. Cancer Res. *44*: 2259 (1984).

290 Hericourt, J.; Richet, C.: 'Physiologie Pathologique' – de la serotherapie dans le traitement du cancer. C. r. hebd. Séanc. Acad. Sci., Paris *121*: 567 (1985).

291 Herlyn, M.; Steplewski, Z.; Herlyn, D.; Koprowski, H.: Colorectal carcinoma specific antigen: detection by means of monoclonal antibodies. Proc. natn. Acad. Sci. USA *76*: 1438 (1979).

292 Herlyn, D.; Koprowski, H.: IgG2a monoclonal antibodies inhibit human tumor growth through interaction with effector cells. Proc. natn. Acad. Sci. USA 79: 4761 (1982).
293 Herlyn, M.; Sears, H.F.; Steplewski, Z.; Koprowski, H.: Monoclonal antibody detection of a circulating tumor associated antigen: I. Presence of antigen in sera patients with colorectal, gastric and pancreatic carcinoma. J. clin. Immunol. 2: 135 (1982).
294 Herlyn, D.; Power, J.; Alavi, A.; Mattis, J.A.; Herlyn, M.; Ernst, C.; Vaum, R.; Koprowski, H.: Radioimmunodetection of human tumor xenografts by monoclonal antibodies. Cancer Res. 43: 2731 (1983).
295 Herlyn, D.; Powe, J.; Ross, A.H.; Herlyn, M.; Koprowski, H.: Inhibition of human tumor growth by IgG2a monoclonal antibodies correlates with antibody density on tumor cells. J. Immunol. 134: 1300 (1985).
296 Herlyn, D.; Lubeck, M.; Sears, H.; Koprowski, H.: Specific detection of anti-idiotypic immune responses in cancer patients treated with murine monoclonal antibody. J. Immunol. Methods 85: 27 (1985).
297 Herlyn, D.; Herlyn, M.; Steplewski, Z.; Koprowski, H.: Monoclonal anti-human tumor antibodies of six isotypes in cytotoxic reactions with human and murine effector cells. Cell. Immunol. 92: 105 (1985).
298 Herlyn, D.; Ross, A.H.; Koprowski, H.: Anti-idiotypic antibodies bear the internal image of a human tumor antigen. Science 232: 100 (1986).
299 Hersey, P.; Schibeci, S.D.; Townsend, P.; Burns, C.; Cheresh, D.A.: Potentiation of lymphocyte responses by monoclonal antibodies to GD_3. Cancer Res. 46: 6083 (1986).
300 Herve, P.: Depletion of T-lymphocytes in donor marrow with Pan-T monoclonal antibodies and complement for prevention of acute graft-versus-host disease: a pilot study on 29 patients. J. natn. Cancer Inst. 76: 1311 (1986).
301 Hinuma, Y.; Nagata, K.; Hanaoka, M.; Nakai, M.; Maisumoto, T.; Kinoshita, T.; Shirakawa, S.; Miyoshi, I.: Adult T-cell leukemia: Antigen in an ATL cell line and detection of antibodies to the antigen in human sera. Proc. natn. Acad. Sci. USA 78: 6476 (1981).
302 Hirohashi, S.; Shimosato, Y.; Ino, Y.: Antibodies from EB-virus-transformed lymphocytes of lymph nodes adjoining lung cancer. Br. J. Cancer 46: 802 (1982).
303 Hirohashi, S.; Shimosato, Y.; Ino, Y.: The in vitro production of tumour-related human monoclonal antibody and its immunohistochemical screening with autologous tissue. Gann 73: 345 (1982).
304 Hirohashi, S.; Ino, Y.; Kodama, T.; Shimosato, Y.: Distribution of blood group antigens A, B, H, and I (MA) in mucus-producing adenocarcinoma of human lung. J. natn. Cancer Inst. 72: 1299 (1984).
305 Hirohashi, S.; Clausen, H.; Nudelman, E.; Inoue, H.; Shimosato, Y., Hakomori, S.: A human monoclonal antibody directed to blood group i antigen: Heterohybridoma between human lymphocytes from regional lymph nodes from a lung cancer patient and mouse myeloma. J. Immunol. 136: 4163 (1986).
306 Hnatowich, D.J.; Childs, R.L.; Lanteigne, D.; Najavi, A.: The preparation of DTPA-coupled antibodies radiolabelled with metallic radionuclides: an improved method. J. Immunol. Methods 65: 147 (1983).
307 Hnatowich, D.J.; Griffin, T.W.; Kosciuczyk, C.; Rusckowski, M.; Childs, R.L.; Mattis, J.A.; Shealy, D.; Doherty, P.W.: Pharmacokinetics of an Indium-111-labeled monoclonal antibody in cancer patients. J. nucl. Med. 26: 849 (1985).

308 Ho, M.-K.; Rand, N.; Murray, J.; Kato, K.; Rabin, H.: In vitro immunisation of human lymphocytes. 1. Production of human monoclonal antibodies against bombesin and tetanus toxoid. J. Immunol. *135*: 3831 (1985).

309 Honjo, T.; Kataoka, T.: Organization of immunoglobulin heavy chain genes and allelic deletion model. Proc. natn. Acad. Sci. USA *75*: 2140 (1978).

310 Horan Hand, P.; Nuti, M.; Colcher, D.; Schlom, J.: Definition of antigenic heterogeneity and modulation among human mammary carcinoma cell populations using monoclonal antibodies to tumor associated antigens. Cancer Res. *43*: 728 (1983).

311 Horan Hand, P.; Colcher, D.; Salomon, D.; Ridge, J.; Noguchi, P.; Schlom, J.: Influence of spatial configuration of carcinoma cell population on the expression of a tumor-associated glycoprotein. Cancer Res. *45*: 833 (1985).

312 Hough, V.W.; Eddy, R.P.; Hamblin, T.J.; Stevenson, F.K.; Stevenson, G.T.: Anti-idiotype sera raised against surface immunoglobulin of human neoplastic lymphocytes. J. exp. Med. *144*: 960 (1976).

313 Houghton, A.N.; Mintzer, D.; Cordon-Cardo, C.; Welte, S.; Fliegel, B.; Vadhan, S.; Carswell, E.; Melamed, M.R.; Oettgen, H.F.; Old, L.J.: Mouse monoclonal IgG3 antibody detecting GD_3 ganglioside: A phase I trial in patients with malignant melanoma. Proc. natn. Acad. Sci. USA *82*: 1242 (1985).

314 Huang, L.C.; Brockhaus, M.; Magnani, J.L.; Cuttitta, S.R.; Minna, J.D.; Ginsburg, V.: Many monoclonal antibodies with an apparent specificity for certain lung cancer are directed against a sugar sequence found in lacto-N-fucopentaose III. Arch. biochem. Biophys. *220*: 318 (1983).

315 Humm, J.L.: Dosimetric aspects of radiolabeled antibodies for tumor therapy. J. nucl. Med. *27*: 1490 (1986).

316 Hurwitz, E.; Levy, R.; Maron, R.; Wilchek, M.; Arnon, R.; Sela, M.: The covalent binding of daunomycin and adriamycin to antibodies, with retention to both drug and antibody activities. Cancer Res. *35*: 1175 (1975).

317 Hurwitz, E.; Maron, R.; Wilchek, M.; Sela, M.: Daunomycin-immunoglobulin conjugates, uptake and activity in vitro. Eur. J. Cancer *14*: 1213 (1978).

318 Hurwitz, E.; Wilchek, M.; Pitha, J.: Soluble macromolecules as carriers for daunorubicin. J. appl. Biochem. *2*: 25 (1980).

319 Hurwitz, E.; Kashi, R.; Burowsky, D.; Arnon, R.; Haimovich, J.: Site-directed chemotherapy with a drug bound to anti-idiotypic antibody to a lymphoma cell-surface IgM. Int. J. Cancer *31*: 745 (1983).

320 Hurwitz, E.; Kashi, R.; Arnon, R.; Wilchek, M.; Sela, M.: The covalent linking of two nucleotide analogues to antibodies. J. med. Chem. *28*: 137 (1985).

321 Hwang, K.; Foon, K.; Cheung, P.; Pearson, J.; Oldham, R.: Selective anti-tumor effect of L10 hepatocarcinoma cells of a potent immunoconjugate composed of the A chain of abrin and a monoclonal antibody to a hepatoma-associated antigen. Cancer Res. *44*: 4578 (1984).

322 Hwang, K.M.; Fodstad, Ö.; Oldham, R.K.; Morgan, A.C., Jr.: Radiolocalization of xenografted human malignant melanoma by a monoclonal antibody (9.2.27) to a melanoma-associated antigen in nude mice. Cancer Res. 45: 4150 (1985).

323 Hylarides, M.; Jones, D.; Seubert, J.; Hadley, S.; Wilbur, S.: Synthesis and radioiodination of Jodophenyl conjugates for protein labeling. J. nucl. Med. *28*: 560 (1987).

324 Imai, K.; Nakanishi, T.; Noguchi, T.; Yachi, A.; Ferrone, S.: Selective in vitro toxicity of

purothionin conjugated to the monoclonal antibody 225.285 to a human molecular-weight melanoma-associated antigen. Cancer Immunol. Immunother. *15*: 206 (1983).

325 Imai, K.; Natali, P.G.; Kay, N.E.; Wilson, B.S.; Ferrone, S.: Tissue distribution and molecular profile of a differentiation antigen detected by a monoclonal antibody (345.134S) produced against human melanoma cells. Cancer Immunol. Immunother. *12*: 159 (1982).

326 Irie, R.F.; Sze, L.L.; Saxton, R.E.: Human antibody to OFA-1, a tumour antigen, produced in vitro by Epstein-Barr virus-transformed human B-lymphoid cell lines. Proc. natn. Acad. Sci. USA *79*: 5666 (1982).

327 IUPAC-IUB Commission on Biochemical Nomenclature. Lipids *12*: 455 (1977).

328 Jackson, P.C.; Pitscher, E.M.; Davies, J.O.; Davies, E.R.; Sadowski, C.S.; Staddon, G.E.; Stirrat, G.M.; Sunderland, C.A.: Radionuclide imaging of ovarian tumors with a radiolabelled (^{123}I) monoclonal antibody (NDOG$_2$). Eur. J. nucl. Med. *11*: 22 (1985).

329 Jacobelli, S.; Natoli, V.; Scambia, G. et al.: A monoclonal antibody (ABB) reactive with human breast cancer. Cancer Res. *45*: 4334 (1985).

330 Jaffers, G.J.; Fuller, T.C.; Cosimi, A.B.; Russell, P.S.; Winn, H.J.; Colvin, R.B.: Monoclonal antibody therapy anti-idiotypic and non-anti-idiotypic antibodies to OKT-3 arising despite intense immunosuppression. Transplantation *41*: 572 (1986).

331 James, K.; Bell, G.T.: Human monoclonal antibody production: Current status and future prospects. J. Immunol. Methods *100*: 5 (1987).

332 Jansen, F.K.; Blythman, H.E.; Carriere, D.; Casellas, P.; Gros, O.; Gros, P.; Laurent, J.C.; Paolucci, F.; Pau, B.; Poncelet, P.: Immunotoxins: hybrid molecules combining high specificity and potent cytotoxicity. Immunol. Rev. *62*: 185 (1982).

333 Jansen, J.; Falkenberg, J.H.F.; Stepan, D.E.; LeBien, T.W.: Removal of neoplastic cells from autologous bone marrow grafts with monoclonal antibodies. Semin. Hematol. *21*: 164 (1984).

334 Jerne, N.K.; Roland, J.; Cazenave, P.A.: Recurrent idiotypes and internal images. EMBO J. *1*: 243 (1982).

335 Jerne, N.K.: Towards a network theory of the immune system. Ann. Immunol. (Paris) *125C*: 373 (1974).

336 Johannsson, A.; Ellis, P.H.; Bates, D.L.; Plumb, A.M.; Stanley, C.J.: Enzyme amplification for immunoassays. Detection limit of one hundredth of an attomole. J. immunol. Methods *87*: 7 (1986).

337 Johnson, J.P.; Demmer-Dieckmann, M.; Meo, T.; Hadam, M.R.; Riethmüller, G.: Surface antigens of human melanoma characterization of two antigens found on cell lines and fresh tumors of diverse tissue origin. Eur. J. Immunol. *11*: 825 (1981).

338 Johnson, J.R.; Ford, C.H.J.; Newman, C.E.; Woodhouse, C.S.; Rowland, G.F.; Simmonds, R.G.: A vindesine-anti-CEA conjugate cytotoxic for human cancer cells in vitro. Br. J. Cancer *44*: 372 (1981).

339 Jones, D.H.; Goldman, A.; Gordon, L.; Pritchard, J.; Gregory, B.J.; Kemshead, J.T.: Therapeutic application of a radiolabeled monoclonal antibody in nude mice xenografted with human neuroblastoma: tumoricidal effects and distribution studies. Int. J. Cancer *35*: 715 (1985).

340 Jones, D.H.; Looney, R.J.; Anderson, C.L.: Two distinct classes of IgG Fc receptors on a human monocyte line (U937) defined by differences in binding of murine IgG subclasses at low ionic strength. J. Immunol. *135*: 3348 (1985).

341 Jones, P.T.; Dear, P.H.; Foote, J.; Neuberger, M.S.; Winter, G.: Replacing the

complementarity-determining regions in a human antibody with those from a mouse. Nature *321*: 522 (1986).

342 Kaizer, H.; Stuart, R.K.; Fuller, D.J.; Braine, H.G.; Serai, R.; Colvin, M.; Wheram, M.D.; Santos, G.W.: Autologous bone marrow transplantation in acute leukemia: progress report on phase I study of 4-hydroperoxycyclophos (4-Hc) incubation of marrow prior to cryopreservation. Proc. Am. Soc. Clin. Oncol. *1*: 131 (1982).

343 Kalofonos, H.; Courtney-Luck, N.; Lavender, J.P.; Hooker, G.; Robinson, G.; Snook, D.; Taylor-Papadimitriou, J.; Epenetos, A.A.: Antibody-guided targeting of non-small lung cancer using radiolabelled HMFG1 (Fab')2 fragments. Br. J. Cancer *55*: suppl. VIII, p. 339 (1987).

344 Kaminski, M.S.; Kitamura, K.; Maloney, D.G.; Campbell, M.J.; Levy, R.: Importance of antibody isotype in monoclonal anti-idiotype therapy of a murine B cell lymphoma. A study of hybridoma class switch variants. J. Immunol. *136*: 1123 (1986).

345 Kanazawa, I.; Yamakawa, T.: Presence of glucosyl ceramide and lactosyl ceramide in human intracranial tumors. Jap. J. exp. Med. *44*: 379 (1974).

346 Kanellos, J.; Pietersz, G.A.; McKenzie, I.F.C.: Studies of methotrexate-monoclonal antibody conjugates for immunotherapy. J. natn. Cancer Inst. *75*: 319 (1985).

347 Kan-Mitchell, J.; Andrews, K.L.; Gallardo, D.; Mitchell, M.S.: Altered antigenicity of human monoclonal antibodies derived from human-mouse heterohybridomas. Hybridoma *6*: 161 (1987).

348 Kannagi, R.; Levine, P.; Watanabe, K.; Hakomori, S.: Glycolipid and glycoprotein profiles and characterization of the major glycolipid antigen in gastric cancer of the 1951 patient of blood group genotype pp (Mrs. D.J.). Cancer Res. *42*: 5249 (1982).

349 Kannagi, R.; Levery, S.B.; Ishigami, F.; Hakomori, S.; Shevinsky, L.H.; Knowles, B.B.; Solter, D.: New globoseries glycosphingolipids in human teratocarcinoma reactive with the monoclonal antibody directed to a developmentally regulated antigen, stage-specific embryonic antigen 3. J. biol. Chem. *258*: 8934 (1983).

350 Kannagi, R.; Stroup, R.; Cochran, N.A.; Urdal, D.L.; Young, W.W., Jr.; Hakomori, S.: Factors affecting expression of glycolipid tumor antigens: influence of ceramide composition and coexisting glycolipid on the antigenicity of gangliotriaosylceramide in murine lymphoma cells. Cancer Res. *43*: 4997 (1983).

351 Karlsson, K.A.; Samuelsson, B.E.; Schersten, T.; Steen, G.O.; Wahlqvist, L.: The sphingolipid composition of human renal carcinoma. Biochim. biophys. Acta *337*: 349 (1974).

352 Kato, Y.; Tsukada, Y.; Hara, T.; Hirai, H.: Enhanced antitumor activity of mitomycin C conjugated with anti-alpha-fetoprotein antibody by a novel method of conjugation. J. Appl. Biochem. *5*: 313 (1983).

353 Kato, Y.; Saito, M.; Fukushima, H.; Takeda, Y.; Hara, T.: Antitumor activitiy of 1-B-D-arabinofuranosylcytosine conjugated with poly-L-glutamic acid and its derivatives. Cancer Res.: 44: 25 (1984).

354 Kato, Y.; Umemoto, N.; Kayama, Y.; Fukushima, H.; Takeda, Y.; Hara, T.; Tsukada, Y.: A novel method of conjugation of daunomycin with antibody with a poly-L-glutamic acid derivative as intermediate drug carrier. An anti-alpha-fetoprotein antibody-daunomycin conjugate. J. med. Chem. *27*: 1602 (1984).

355 Kennedy, J.L.; Adler-Storthz, K.; Henkel, R.D.; Sanchez, Y.; Melnick. J.L.; Dreesman, G.R.: Immune response to hepatitis B surface antigen: enhancement by prior injection of antibodies to the idiotype. Science *221*: 853 (1983).

356 Kennedy, R.C.; Dreesman, G.R.: Enhancement of the immune response to hepatitis B surface antigen: In vivo administration of antiidiotype induces anti-HBs that expresses a similar idiotype. J. exp. Med. *159*: 655 (1984).

357 Kennedy, R.C.; Dreesman, G.R.; Kohler, H.: Vaccines utilizing internal image of anti-idiotypic antibodies that mimic antigens of infectious organisms. Biotechniques *3*: 404 (1985).

358 Kennel. S.J.: Binding of monoclonal antibody to protein antigen in fluid phase or bound to solid supports. J. immunol. Methods *55*: 1 (1982).

359 Kennel, S.J.; Foote, L.J.; Lankford, P.K.; Johnson, M.; Mitchell, T.; Braslawsky, G.R.: Direct binding or radioiodinated monoclonal antibody to tumor cells: significance of antibody purity and affinity for drug targeting or tumor imaging. Hybridoma *2*: 297 (1983).

360 Kern, H.F.; Bosslet, K.; Mollenhauer, J.; Sedlacek, H.H.; Schorlemmer, H.U.: Monocyte-related functions expressed in cell lines established from human pancreatic adenocarcinoma. I. Comparative analysis of endocytotic activity, lysosomal enzyme secretion, and superoxide anion production. Pancreas *2*: 212 (1987).

361 Khaw, B.A.; Cooney, J.; Edgington, T.; Strauss, H.W.: Differences in experimental tumor localization of dual-labeled monoclonal antibody. J. nucl. Med. *27*: 1293 (1986).

362 Kimura, I.; Ohnoshi, T.; Tsubota, T.; Kobayashi, T.; Abe, S.: Production of tumor antibody-neocarcinostatin (NCS) conjugate and its biological activity. Cancer Immunol. Immunother. *7*: 235 (1980).

363 Kimura, I.; Tsubota, T.; Ohnoshi, T.; Sato, Y.; Okazaki, M.; Manabe, Y.; Abe, S.: In vivo antitumor activity of neocarcinostatin (NCS)-tumor antibody conjugate against a transplantable human leukemia cell line (BALL-1). Jap. J. clin. Oncol. *13*: 425 (1983).

364 King, T.P.; Li, Y.; Kochoumian, L.: Preparation of protein conjugates via intermolecular disulfide bond formation. Biochemistry *17*: 1499 (1978).

365 Kipps, T.J.; Parham, P.; Punt, J.; Herzenberg, L.A.: Importance of immunoglobulin isotype in human antibody dependent, cell-mediated cytotoxicity directed by murine monoclonal antibodies. J. exp. Med. *161*: 1 (1985).

366 Klein, G.; Sjörgren, H.O.; Klein, E.; Hellström, K.E.: Demonstration of resistance against methylcholanthrene-induced sarcomas in the primary authochthonous host. Cancer Res. *20*: 1561 (1960).

367 Kleist, S. von, Chavanel, G.; Burtin, P.: Identification of an antigen from normal human tissue that cross-reacts with the carcinoembryonic antigen. Proc. natn. Acad. Sci. USA *69*: 2492 (1972).

368 Klotz, I.M.; Heiney, R.E.: Introduction of sulfhydryl groups into proteins using acetylmercaptosuccinic anhydride. Archs. Biochem. Biophys. *96*: 605 (1962).

369 Kniep, B.; Monner, D.A.; Burrichter, H.; Diehl, V.; Muhlradt, P.F.: Gangliotriaosylceramide (asialo G_{M2}), a glycosphingolipid marker for cell lines from patients with Hodgkin's disease. J. Immunol. *131*: 1591 (1983).

370 Köhler, G.; Milstein, C.: Continuous cultures of fused cells secreting antibody of predefined specificity. Nature *256*: 495 (1975).

371 Kohler, H.: The immune network revisited; in Kohler, Urbain, Cazenave, Idiotype in biology and medicine, p. 3 (Academic Press New York 1984).

372 Koprowski, H.; Herlyn, M.; Steplewski, Z.; Sears, H.F.: Specific antigen in serum of patients with colon carcinoma. Science *212*: 53 (1981).

373 Koprowski, H.; Herlyn, D.; Lubeck, M.; De Freitas, E.; Sears, H.F.: Human anti-idiotype antibodies in cancer patients: Is the modulation of the immune response beneficial for the patient? Proc. natn. Acad. Sci. USA *81*: 216 (1984).

374 Korbling, M.; Hess, A.D.; Tutschka, P.J.; Kaizer, H.; Colvin, M.D.; Santos, G.W.: 4-Hydroperoxycyclophosphamide, a model for eliminating residual human tumor cells and T-lymphocytes from the bone marrow graft. Br. J. Haemat. 52: 89 (1982).
375 Korngold, L.; Pressman, D.: The localization of antilymphosarcoma antibodies in the Murphy lymphosarcoma of the rat. Cancer Res. 14: 96 (1954).
376 Kozbor, D.; Roder, J.C.: The production of monoclonal antibodies from human lymphocytes. Immunol. Today 4: 72 (1983).
377 Kozbor, D.; Roder, J.C.: In vitro stimulated lymphocytes as a source of human hybridomas. Eur. J. Immunol. 14: 23 (1984).
378 Kozbor, D.; Lagarde, A.E.; Roder, J.C.: Human hybridomas constructed with antigen-specific Epstein-Barr virus-transformed cell lines. Proc. natn. Acad. Sci. USA 79: 6651 (1982).
379 Kozbor, D.; Dexter, D.; Roder, J.C.: A comparative analysis of the phenotype characteristics of available fusion partners for the construction of human hybridomas. Hybridoma 2: 7 (1983).
380 Kozbor, D.; Croce, C.M.: Fusion partners for production of human monoclonal antibodies; in Engelman, Foung, Larrick, Raubitschek, Human hybridomas and monoclonal antibodies, p. 21 (Plenum Press, New York 1985).
381 Kozbor, D.; Abramow-Newerly, W.; Tripputi, P.; Cole, S.P.C.; Weibel, J.; Roder, J.C.; Croce, C.M.: Specific immunoglobulin production and enhanced tumorigenicity following ascites growth of human hybridoma. J. immunol. Methods 81: 31 (1985).
382 Kraemer, H.P.; Sedlacek, H.H.: Human tumor test systems: A new screening approach. Behring Inst. Mitt. 80: 102 (1986).
383 Krolick, K.A.; Uhr, J.W.; Vitetta, E.S.: Selective killing of leukemia cells by antibody-toxin conjugates: implications for autologous bone marrow transplantation. Nature (London) 295: 604 (1982).
384 Krolick, K.; Uhr, J.W.; Slavin, S.; Vitetta, E.S.: In vivo therapy of a murine B cell tumor (BCL_1) using antibody-ricin A chain immunotoxins. J. exp. Med. 155: 1797 (1982).
385 Kufe, D.; Inghirami, G.; Abe, M.; Hayes, D.; Justi-Wheeler, H.; Schlom, J.: Differential reactivity of a novel monoclonal antibody (DF3) with human malignant versus benign breast tumors. Hybridoma 3: 233 (1984).
386 Kularni, P.N.; Blair, A.H.; Ghose, T.: Covalent binding of methotrexate to immunoglobulins and the effect of antibody-linked drug on tumor growth in vivo. Cancer Res. 41: 2700 (1981).
387 (See reference no. 386).
388 Kulkarni, P.N.; Blair, A.H.; Ghose, T.; Mammen, M.: Conjugation of methotrexate to IgG antibodies and their $F(ab)_2$ fragments and the effect of conjugated methotrexate on tumor growth in vivo. Cancer Immunol. Immunother. 19: 211 (1985).
389 Kuroki, M.; Shinoda, T.; Takayasu, T.; Koga, Y.; Matsuoka, Y.: Immunological characterization and structural studies of normal fecal antigen-1 related to carcinoembryonic antigen. Mol. Immunol. 19: 399 (1982).
390 Kuroki, M.; Koga, Y.; Matsuoka, Y.: Purification and characterization of carcinoembryonic antigen-related antigens in normal adult feces. Cancer Res. 41: 713 (1981).
391 Kuroki, M.; Arakawa, F.; Higuchi, H.; Matsunaga, A.; Okamoto, N.; Takakura, K.; Matsuoka, Y.: Epitope mapping of the carcinoembryonic antigen by monoclonal antibodies and establishment of a new improved radioimmunoassay system. Jap. J. Cancer Res. 78: 386 (1987).

392 Kurth, R.; Fenyö, E.M., Klein, E.; Essex, M.: Cell-surface antigens induced by RNA tumour viruses. Nature 279: 197 (1979).
393 Kurtzman, S.H.; Russo, A.; Mitchell, J.B.; DeGraff, W.; Sindelar, W.F.; Brechbiel, M.W.; Gansow, O.A.; Friedman, A.M.; Hines, J.J.; Gamson, J.; Atcher, R.W.: ^{212}Bismuth linked to an anti-pancreatic carcinoma antibody: a model for alpha particle-emitter radioimmunotherapy. (in press).
394 Kyoizumi, S.; Akiyma, M.; Kouno, N.; Kobuke, K.; Hakoda, M.; Jones, S.L.; Yamakido, M.: Monoclonal antibodies to human sqamous cell carcinoma of the lung and their application to tumor diagnosis. Cancer Res. 45: 3274 (1985).
395 Lamerz, R.; Fateh-Moghadam, A.: Carcinofetale Antigene: III. Andere carcinofetale Antigene. Klin. Wschr. 53: 403 (1975).
396 Lamerz, R.: Fateh-Moghadam, A.: Carcinofetale Antigene: II. Carcinoembryonales Antigen (CEA). Klin. Wschr. 53: 193 (1975).
397 Larson, R.W.; Ferens, J.M.; Graham, M.M.; Hill, L.D.; Beaumier, P.L.: Localization of 131-labeled P97-specific Fab fragments in human melanoma as a basis for radiotherapy. J. clin. Invest. 72: 2101 (1983).
398 Larson, S.M.; Brown, J.P.; Wright, P.W.; Carrasquillo, J.A.; Hellstrom, I.; Hellstrom K.E.: Imaging of melanoma with ^{131}I-labelled monoclonal antibodies. J. nucl. Med. 24: 123 (1983).
399 Larson, S.M.; Carrasquillo, J.A.; McGuffin, R.W.; Krohn, K.A.; Ferens, J.M.; Hill, L.D.; Beaumier, P.L.; Reynolds, J.C.; Hellstrom, K.E.; Hellstrom, I.: Use of ^{131}I labelled immune Fab against high molecular weight antigen of human melanoma: Preliminary experience. Radiology 155: 487 (1985).
400 Larson, S.M.; Carrasquillo, J.A.; Reynolds, J.C.; Hellstrom, I.; Hellstrom, K.-E.; Mulshine, J.C.; Mattis, L.E.: Therapeutic applications of radiolabelled antibodies: current situation and prospects. Nucl. Med. Biol. 13: 207 (1986).
401 Lashford, L.; Jones, D.; Pritchard, J.; Gordon, J.; Breatnach, F.; Kemshead, J.T.: Therapeutic application of radiolabelled monoclonal antibody UJ13A in children with disseminated neuroblastoma (Abstract). Cancer Drug Delivery 2: 233 (1985).
402 Laszlo, J.; Buckley, C.E.; Amos, C.B.: Infusion of isologous immune plasma in chronic lymphocytic leukemia. Blood 31: 104 (1986).
403 Lee, F.H.; Berczi, I.; Fujimoto, S.; Sehon, A.H.: The use of antifibrin antibodies for the destruction of tumor cells. III. Complete regression of MC-D sarcoma in guinea-pigs by conjugates of daunomycin with antifibrin antibodies. Cancer Immunol. Immunother. 5: 201 (1978).
404 Lee, W.M.; Westrick, M.A.; Macher, B.A.: Neutral glycosphingolipids of human acute leukemias. J. biol. Chem. 257: 10090 (1982).
405 Lee, V.K.; Harriott, T.G.; Kuchroo, V.J.; Halliday, W.J.; Hellstrom, I.; Hellstrom, K.E.: Monoclonal anti-idiotypic antibodies related to a murine oncofetal bladder tumor antigen induce specific cell-mediated tumor immunity. Proc. natn. Acad. Sci. USA 82: 6286 (1985).
406 Lee, Y., Bullard, D.E.; Wikstrand, C.J.; Zalutsky, M.R.; Muhlbaier, L.H.; Bigner, D.D.: Comparison of monoclonal antibody delivery to intracranial glioma xenografts by intravenous and intracarotid administration. Cancer Res. 47: 1941 (1987).
407 Leichner, P.K.; Klein, J.L.; Garrison, J.B.; Jenkins, R.E.; Nickoloff, E.L.: Dosimetry of ^{131}I-labelled anti-ferritin in hepatoma: A model for radioimmunoglobulin dosimetry. Int. J. Radiat. Oncol. Biol. Phys. 7: 323 (1981).

408 Leichner, P.K.; Klein, J.L.; Siegelman, S.S.; Ettinger, D.; Order, S.E.: Dosimetry of ^{131}I-labelled anti-ferritin in hepatoma: Specific activities in the tumor and liver. Cancer Treat. Rep. 67: 647 (1983).

409 Lemkin, S.; Tikita, K.; Sherman, G.; Simko, T.; Schwartz, L.; Ciccirilli, J.; Drew, S.; Terasaki, P.: Phase I-II study of monoclonal antibodies in gastrointestinal cancer, abstracted. Proc. Am. Soc. Clin. Oncol. 3: 47 (1984).

410 Lenhard, R.E.; Order, S.E.; Spunberg, J.J.; Asbell, S.O.; Leibel, S.A.: Isotopic immunoglobulin: A new systemic therapy for advanced Hodgkin's disease. J. clin. Oncol. 3: 1296 (1985).

411 Leonard, J.E.; Taetle, R.; To, D.; Rhyner, K.: Preclinical studies on the use of selective antibody-ricin conjugates in autologous bone marrow transplantation. Blood 65: 1149 (1985).

412 Leung, C.S.H.; Meares, C.F.; Goodwin, D.A.: The attachment of metal-chelating groups to proteins: Tagging of albumin by diazonium coupling and use of the products as radiopharmaceuticals. J. appl. Radiat. Isot. 29: 687 (1978).

413 Levine, P.; Bobbit, O.B.; Waller, R.K.; Kuhmichel, A.: Isoimmunization by a new blood factor in tumor cells. Proc. Soc. exp. Biol. Med. 777: 403 (1951).

414 Levy, R.L.; Miller, R.A.: Biological and clinical implications of lymphocyte hybridomas: tumor therapy with monoclonal antibodies. Ann. Rev. Med. 34: 107 (1983).

415 Lindmo, T.; Boven, E.: Cuttitta; F.; Fedorko, J.; Bunn, P.A., Jr.: Determination of the immunoreactive fraction of radiolabeled monoclonal antibodies by linear extrapolation to binding at infinite antigen excess. J. Immunol. Methods 72: 77 (1984).

416 Linford, J.H.; Froese, G.; Berczi, I.; Israels, L.G.: An alkylating agent-globulin conjugate with both alkylating and antibody activity. J. natn. Cancer Inst. 52: 1665 (1974).

417 Linz, U.; Stöcklin, E.: Chemical and biological consequences of the radiation decay of iodine-125 in plasmid DNA. Radiation Res. 101: 262 (1985).

418 Little, C.D.; Nau, M.M.; Carney, D.N.; Gazdar, A.F.; Minna, J.D.: Amplification and expression of the c-myc oncogene in human lung cancer cell lines. Nature 306: 194 (1983).

419 Liu, F.T.; Zinnecker, M.; Hamaoka, T.; Katz, D.H.: New procedures for preparation and isolation of conjugates of proteins and a synthetic copolymer of D-amino acids and immunochemical characterization of such conjugates. Biochemistry 18: 690 (1979).

420 Liu, A.Y.; Robinson, R.R.; Hellström, K.E.; Murray, E.D., Jr.; Chang, C.P.; Hellström, I.: Chimeric mouse-human IgG1 antibody that can mediate lysis of cancer cells. Proc. natn. Acad. Sci. USA 84: 3439 (1987).

421 Lobuglio, A.F.; Saleh, M.; Peterson, L.; Wheeler, R.; Carrano, R.; Huster, W.E.; Khazaeli, M.B.: Phase I clinical trial of CO17-1A monoclonal antibody. Hybridoma 5: suppl. 1, 117 (1986).

422 Lowder, J.N.; Meeker, T.C.; Campbell, M.; Garcia, C.F.; Gralow, J.; Miller, R.A.; Warnke, R.; Levy, R.: Studies on B-lymphoid tumors treated with monoclonal anti-idiotypic antibodies: correlation with clinical responses. Blood 69: 199 (1987).

423 Lowder, J.N.: The current status of monoclonal antibodies in the diagnosis and therapy of cancer. Curr. Probl. Cancer 10: 485 (1986).

424 Mach, J.-P.; Pusztaszeri, G.: Carcinoembryonic antigen (CEA): demonstration of a partial identity between CEA and a normal glycoprotein. Immunochemistry 9: 1031 (1972).

425 Mach, J.-P.; Carrel, S.; Merenda, C.; Sordat, D.; Cerottini, J.C.: In vivo localisation of radiolabelled antibodies to carcinoembryonic antigen in human colon carcinoma grafted into nude mice. Nature 248: 704 (1974).

426 Mach, J.-P.; Carrel, S.; Forni, M.; Ritschard, J.; Donath, A.; Alberto, P.: Tumor localization of radiolabeled antibodies against carcinoembryonic antigen in patients with carcinoma. A critical evaluation. New Engl. J. Med. *303*: 5 (1980).

427 Mach, J.-P.; Buchegger, F.; Forni, M.; Ritschard, J.; Bersche, C. Lumbroso, J. D.; Schreyer, M.; Giradet, C.; Accola, R. S.; Carrel, S.: Use of radiolabelled monoclonal anti-CEA antibodies for the detection of human carcinomas by external photoscanning and tomoscintigraphy. Immunol. Today *2*: 239 (1981).

428 (See reference no. 427).

429 Mach, J.-P.; Chatal, J. F.; Lumbroso, J. D.; Buchegger, F.; Forni, M.; Ritschard, J.; Berche, C.; Douillard, J. Y.; Carrel, S.; Herlyn, M.; Steplewski, Z.; Koprowski, H.: Tumor localization in patients by radiolabeled monoclonal antibodies against colon cancer. Cancer. Res. *43*: 5593 (1983).

430 Mach, J.-P.; Carrel, S.; Forni, M.; Ritschard, J.; Donath, A.; Alberto, P.: Tumor localization of radiolabelled antibodies against carcinoembryonic antigen in patients with carcinoma. Cancer Res. *43*: 5593 (1983).

431 Mach, J.-P.; Pettawel, J.; Bishos-De-La-Loi, A.; De La Loi, B.: The radiolabelled monoclonal anti-CEA antibodies and fragments in colon carcinoma, diagnostic and therapeutic approach. Proc. 'Advances in the applications of monoclonal antibodies in clinical oncology', p. 20, London (1987).

432 Magnani, J. L.; Brockhaus, M.; Smith, D. F.; Ginsburg, V.; Blaszczyk, M.; Mitchell, K. F.; Steplewki, Z.; Koprowski, H.: A monosialoganglioside is a monoclonal antibody-defined antigen of colon carcinoma. Science *212*: 55 (1981).

433 Magnani, J. L.; Nilsson, B.; Brockhaus, M.; Zopr, D.; Steplewski, Z.; Koprowski, H.; Ginsburg, V.: A monoclonal antibody-defined associated with gastrointestinal cancer is a ganglioside containing sialylated lacto-N-fucopentaose II. J. biol. Chem. *257*: 14365 (1982).

434 Major, P. P.; Kovac, P. E.; Lavallée, M. L.: Monoclonal antibodies to antigens abnormally expressed in breast cancer. J. Histochem. Cytochem. *35*: 139 (1987).

435 Malesci, A.; Tommasini, M. A.; Bocchia, P.; Zerbi, A.; Beretta, E.; Vecchi, M.; Di Carlo, V.: Differential diagnosis of pancreatic cancer and chronic pancreatitis by a monoclonal antibody detecting a new cancer-associated antigen (CA 19-9). La Ricerca Clin. Lab. *14*: 303 (1984).

436 Manabe, Y.; Tsubota, T.; Haruta, Y.; Okazaki, M.; Haisa, S.; Nakamura, K.; Kimura, I.: Production of a monoclonal antibody-bleomycin conjugate utilizing dextran-T40 and the antigen-targeting cytotoxicity of the conjugate. Biochem. biophys. Res. Commun. *115*: 1009 (1983).

437 Manabe, Y.; Tsubota, T.; Haruta, Y.; Kataoka, K.; Okazaki, M.; Haisa, S.; Nakamura, K.; Kimura, I.: Production of a monoclonal antibody-methotrexate conjugate utilizing dextran-T40 and its biological activity. J. Lab. clin. Med. *104*: 445 (1984).

438 Manabe, Y.; Tsubota, T.; Haruta, Y.; Ktaoka, K.; Okazaki, M.; Haisa, S.; Nakamura, K.; Kimura, I.: Production of a monoclonal antibody-mitomycin C conjugate utilizing dextran-T40, and its biological activity. Biochem. Parmac. *34*: 289 (1985).

439 Mann, B. D.; Cohen, M. B.; Saxton, R. E.; Morton, D. L.; Benedict, W. F.; Korn, E. L.; Spolter, L.; Graham, L. S.; Chang, C. C.; Burk, M. W.: Imaging of human tumor xenografts in nude mice with radiolabeled monoclonal antibodies. Cancer *54*: 1318 (1984).

440 Marangos, P. J.; Gazdar, A. F.; Carney, D. N.: Neuron specific enolase in human small cell carcinoma cultures. Cancer lett. *15*: 67 (1982).

441 Martin, P.J.; Astley, C.; Giorgio, A.; Weiblen, B.; Valeri, C.R.; Hansen, J.A.: Observations on the in vivo fate of cells coated with murine monoclonal antibodies; in Reisfeld, Sell, Monoclonal antibodies in cancer therapy, UCLA Symposia, vol. 27, p. 121. (Alan Arvis, New York 1986).
442 Masi, R.; Presciullesi, E.; Ferri, P. et al.: Immunodetection of human melanoma metastases by means of F(ab')$_2$ fragments of monoclonal antibodies: The usefulness of digital images; in Donato, Britton, Immunoscintigraphy, p. 383 (Gordon and Breach, New York 1985).
443 Masuho, Y.; Hara, T.; Noguchi, T.: Preparation of a hybrid of fragment Fab' of antibody and fragment A of diphtheria toxin and its cytotoxicity. Biochem. biophys. Res. Commun. *90:* 320 (1979).
444 Masui, H.; Moroyama, T.; Mendelsohn, J.: Mechanism of antitumor activity in mice for antiepidermal growth factor receptor monoclonal antibodies with different isotypes. Cancer Res. *46:* 5592 (1986).
445 Mathé, G.; Loc, T.; Bernard, J.: Effect sur la leucémie 1210 de la souris d'un combinaison par diazotation d'A-methoptérine et de γ-globulines de hamsters porteurs de cell leucémie par hétérograffe. C. Acad. Sci. (Paris) *246:* 1626 (1958).
446 Matsushita, S.; Robert-Gurhoff, M.; Trepel, J.; Cossman, J.; Mitsuya, H.; Broder, S.: Human monoclonal antibodies directed against an envelope glycoprotein of human T-cell leukemia virus type 1. Proc. natn. Acad. Sci. USA *83:* 2672 (1986).
447 Matzku, S.; Kirchgessner, H.; Dippold, W.G.; Brüggen, J.: Immunoreactivity of monoclonal anti-melanoma antibodies in relation to the amount of radioactive iodine substituted to the antibody molecule. Eur. J. nucl. Med. *11:* 260 (1985).
448 McNamara, M.K.; Ward, R.E.; Kohler, H.: Monoclonal idiotype vaccine, against Streptococcus pneumoniae infection. Science *226:* 1325 (1984).
449 Meares, C.F.; McCall, M.J.; Reardan, D.T.; Goodwin, D.A.; Diamanti, C.I.; McTigue, M.: Conjugation of antibodies with bifunctional chelating agents: Isothiocyanate and bromacetamide reagents, methods of analysis and subsequent addition of metal ions. Anal. Biochem. *142:* 68 (1984).
450 Meares, C.F.; Goodwin, D.A.: Linking radiometals to proteins with bifunctional chelating agents. J. Protein Chem. *3:* 215 (1984).
451 Meeker, T.C.; Lowder, J.; Maloney, D.G.; Miller, R.A.; Thielemans, K.; Warnke, R.; Levy, R.: A clinical trial of anti-idiotypic therapy for B-cell malignancy. Blood *65:* 1349 (1985).
452 Melchert, F.; Kreienberg, R.: Tumormarker und ihre Bedeutung in der gynäkologischen Onkologie. Gynäkologie *13:* 74 (1980).
453 Melino, G.: Drug targeting for neuroblastoma: a case report. Protides biol. Fluids *32:* 445 (1985).
454 Melino, G.; Hazarika, P.; Elliott, P.; Hobbs, J.R.; Cooke, K.B.: Protein-bound daunorubicin as an agent for cancer therapy. Biochem. Soc. Trans. *10:* 505 (1982).
455 Melino, G.; Elliott, P.; Cooke, K.B.; Evans, A.; Hobbs, J.R.: Allogeneic antibodies (Abs) for drug targeting to human neuroblastoma (Nb). Proc. Am. Soc. Clin. Oncol. *3:* 47 (1984).
456 Melino, G.: Drug targeting for neuroblastoma: a case report. J. Cell Biochem. *9A:* suppl., p. 445 (1985).
457 Melino, G.; Hobbs, J.R.; Radford, M.; Cooke, K.B.; Evans, A.M.; Castello, M.A.; Forrest, D.M.: Drug targeting for 7 neuroblastoma patients using human polyclonal antibodies. Protides. biol. Fluids *32:* 413 (1985).

458 Merritt, W.D.; Casper, J.T.; Lauer, S.J.; Reaman, G.H.: Expression of GD_3 ganglioside in childhood T-cell lymphoblastic malignancies. Cancer Res. *47:* 1724 (1987).

459 Metzgar, R.S.; Gaillard, M.T.; Levine, S.J.; Tuck, F.L.; Bossen, E.H.; Borowitz, M.J.: Antigens of human pancreatic adenocarcinoma cells defined by murine monoclonal antibodies. Cancer Res. *42:* 601 (1982).

460 Miller, R.A.; Maloney, D.G.; McKillop. J.; Levy, R.: In vivo effects of murine hybridoma monoclonal antibody in a patient with T-cell leukemia. Blood *58:* 78 (1981).

461 Miller, R.A.; Maloney, D.G.; Warnke, R.; Levy, R.: Treatment of B cell lymphoma with monoclonal anti-idiotypic antibody. New Engl. J. Med. *306:* 517 (1982).

462 Miller, R.A.; Oseroff, A.R.; Stratte, P.T.; Levy, R.: Monoclonal antibody therapeutic trials in seven patients with T cell lymphoma. Blood *62:* 988 (1983).

463 Miller, R.A.; Lowder, J.; Meeker, R.C.; Brown, S.; Levy, R.: Anti-idiotypes in B cell tumor therapy. NCI Monogr. *3:* 131 (1987).

464 (See reference no. 463).

465 Moldofsky, P.J.; Powe, J.; Mülhern, C.B.; Hammond, N.; Sears, H.F.; Gatenby, R.A.; Steplewski, Z.; Koprowski, H.: Metastatic colon carcinoma detected with radiolabeled $F(ab')_2$ monoclonal antibody fragments. Radiology *149:* 549 (1983).

466 Monsigny, M.; Kieda, C.; Roche, A.-C.; Delmotte, F.: Preparation and biological properties of a covalent antitumor drugarm-carrier (DAC conjugate). FEBS Lett. *119:* 181 (1980).

467 Montz, R.; Klapdor, R.; Kremer, B.; Rothe, B.: Immunszintigraphie und SPECT bei Patienten mit Pankreaskarzinom. Nucl. Med. *24:* 232 (1985).

468 Montz, R.; Klapdor, R.; Rothe, B.; Heller, M.: Immunoscintigraphy and radioimmunotherapy in patients with pancreatic carcinoma. Nucl. Med. *25:* 239 (1986).

469 Moolten, F.L.; Cooperband, S.R.: Selective destruction of target cells by diphtheria toxin conjugated to antibody directed against antigens on the cells. Science *169:* 68 (1970).

470 Moolten, F.L.; Capparell, N.J.; Cooperband, S.R.: Antitumor effects of antibody-diphtheria toxin conjugates: use of hapten-coated tumor cells as an antigenic target. J. natn. Cancer Inst. *49:* 1057 (1972).

471 Morishima, X.; Sao, H.; Ueda, R.; Morishita, Y.; Murase, T.; Kodera, X.; Ohno, R.; Tahara, T.; Yoshikawa, S.; Kato, Y.; Yokomaku, S.; Takahashi, T.; Saito, H.: Preliminary clinical trial of autologous bone marrow transplantation after in vitro monoclonal antibody and complement treatments in null cell-type acute lymphocytic leukemia. Jap. J. Cancer Res. *76:* 1222 (1985).

472 Morrison, S.L.; Johnson, M.J.; Herzenberg, L.A.; Oi, V.T.: Chimeric human antibody molecules: mouse antigen-binding domains with human constant region domains. Proc. natn. Acad. Sci. USA *81:* 6851 (1984).

473 Moshakis, V.; McIlhanney, R.A.J.; Raghavan, D.; Neville, A.M.: Localization of human tumour xenografts after i.v. administration of radiolabelled monoclonal antibodies. Br. J. Cancer *44:* 91 (1981).

474 (See reference no. 473).

475 Moss, G.P.; Reese, C.B.; Schofield, K.; Shapiro, R.; Todd, L.: J. chem. Soc. (1963) (cited in reference no. 42).

476 Motta-Hennessy, C.; Eccles, S.A.; Dean, C.; Coghlan, G.: Preparation of ^{67}Ga-labelled human IgG and its Fab fragments using desferoxamine as chelating agent. Eur. J. nucl. Med. *11:* 240 (1985).

477 Müller, J.; Pfleiderer, G.: J. appl. Biochem. *1:* 301 (1979).

478 Muhrer, K.H.; Bosslet, K.; Burkhardt, M.; Müller, H.: Intraarterielle Infusion eines monoklonalen Anti-CEA-Antikörpers bei Lebermetastasen kolorektaler Karzinome. Tumor Diagnostik & Therapie 8: 85 (1987).
479 Muirhead, M.; Martin, P.J.; Torok-Storb, B.; Uhr, J.W.; Vitetta, E.: Use of an antibody-ricin A-chain conjugate to delete neoplastic B cells from human bone marrow. Blood 62: 327 (1983).
480 Mujoo, K.; Cheresh, D.A.; Yang, H.M.; Reisfeld, R.A.: Disialoganglioside GD_2 on human neuroblastoma cells: target antigen for monoclonal antibody-mediated cytolysis and suppression of tumor growth. Cancer Res. 47: 1098 (1987).
481 Mulshine, J.L.; Cuttitta, F.; Bibro, M.; Fedorko, J.; Fargion, S.; Little, C.; Carney, D.N.; Gazdar, A.F.; Minna, J.D.: Monoclonal antibodies that distinguish non-small cell from small cell lung cancer. J. Immunol. 131: 497 (1983).
482 Munz, D.L.; Alavi, A.; Koprowski, H.; Herlyn, D.: Improved radioimmunoimaging of human tumor xenografts by a mixture of monoclonal antibody (Fab')$_2$ fragments. J. nucl. Med. 27: 1739 (1986).
483 Munz, D.L.; Alavi, A.; Koprowski, H.; Herlyn, D.: Enhancement of tumor contrast on radioimmunoscans by using mixtures of monoclonal antibody F(ab')$_2$ fragments. Nucl. Med. 25: 216 (1986).
484 Murakami, H.; Hashizume, S.; Ohashi, H.; Shinohara, K.; Yasumoto, K.; Nomoto, K.; Omura, H.: Human-human hybridomas secreting antibodies specific to human lung carcinoma. In Vitro Cell. Devl. Biol. 21: 593 (1985).
485 Muraro, R.; Wunderlich, D.; Thor, A.; Lundy, J.; Noguchi, P.; Cunningham, R.; Schlom, J.: Definition by monoclonal antibodies of a repertoire of epitopes on carcinoembryonic antigen differentially expressed in human colon carcinomas versus normal adult tissues. Cancer Res. 45: 5769 (1985).
486 Murray, G.: Experiments in immunity in cancer. Can. med. Ass. J. 79: 249 (1958).
487 Murray, J.L.; Rosenblum, M.G.; Sobol, R.E.; Bartholomew, R.M.; Plager, C.E.; Haynie, T.P.; Johns, M.F.; Glenn, H.J.; Lamki, L.; Benkamin, R.S.; Papadopoulous, N.; Boddie, A.W.; Frincke, J.M.; David, G.S.; Carol, D.J.; Hersh, E.M.: Radioimmunoimaging in malignant melanoma with ^{111}In-labeled monoclonal antibody 96.5. Cancer Res. 45: 2376 (1985).
488 Murray, J.L.; Rosenblum, L.; Lamki, L.; Haynie, T.P.; Glenn, H.; Jahns, M.; Plager, C.; Hersh, E.M.; Unger, M.; Carlo, D.L.: Imaging findings and pharmacokinetics of ^{111}Indium ZME-018 monoclonal antibody in malignant melanoma. J. Nucl. Med. 26: P16 (1985).
489 Myers, C.D.; Thorpe, P.E.; Ross, W.C.J.; Cumber, A.J.; Katz, F.E.; Tay, W.: Greaves, M.F.: An immunotoxin with therapeutic potential in T cell leukemia: WT1 ricin A. Blood 63: 1178 (1984).
490 Nadler, L.M.; Stashenko, P.; Hardy, D.W.; Antman, K.H.; Schlossman, S.T.: Serotherapy of a patient with a monoclonal antibody directed against a human lymphoma-associated antigen. Cancer Res. 40: 3147 (1980).
491 Najafi, A.; Fawcett, H.D.; Hutchinson, N.: Comparison of two methods of labelling proteins with In-111. Proc. Am. Chem. Soc., Div. of Nucl. Chemistry and Technology, Miami Beach, USA; Abstract No. 52, (May 1985).
492 Najali, A.; Hutchison, N.: The use of diactivated ester in coupling DTPA to proteins. Int. J. Radiat. Appl. Instrum. Part A Appl. Radiat. Isot. 37: 548 (1986).
493 Nakamura, N.; Chopra, I.J.; Solomon, D.H.: Preparation of high-specific-activity radioactive iodothyronines and their analogues. J. nucl. Med. 18: 1112 (1977).

494 Nepom, G.T.; Nelson, K.A.; Holbeck, S.L.; Hellstrom, I.; Hellstrom, K.E.: Induction of immunity to a human tumor marker by in vivo administration of anti-idiotypic antibodies in mice. Proc. natn. Acad. Sci. USA *81:* 2864 (1984).
495 Neuberger, M.S.; Williams, G.T.; Mitchell, E.B.; Jouhal, S.S.; Flanagen, J.G.; Rabbitts, T.H.: A hapten-specific chimaeric IgE antibody with human physiological effector function. Nature *314:* 268 (1985).
496 Neumann, R.D.; Kirkwood, J.M.; Zoghbi, S.S.; Ernstoff, M.S.; Cornelius, E.A.; Hoffer, P.B.; Gottschalk, A.: Ga-67 vs ^{111}In-DTPA-Anti-p97 monoclonal antibody for scintigraphic detection of metastatic melanoma. J. nucl. Med. *26:* P15 (1985).
497 Neville, D.M.; Chang, T.M.: Receptor-mediated protein transport into cells. Entry mechanisms for toxins, hormones, antibodies, viruses, lysosomal hydrolases, asialoglycoproteins, and carrier proteins. Curr. Top. Membr. Transp. *10:* 66 (1978).
498 Nillson, O.; Mansoon, J.-E.; Brezicka, T.; Holmgren, J.; Lindholm, L.; Sorenson, S.; Yngvason, F.; Svennerholm, L.: Fucosyl-GM$_1$ – a ganglioside associated with small cell lung carcinomas. Glycoconjugate J. *1:* 43 (1984).
499 Nilsson, O.; Brezicka, F.T.; Holmgren, J.; Sörenson, S.; Svennerholm, L.; Yngvason, F.; Lindholm, L.: Detection of a ganglioside antigen associated with small cell lung carcinomas using monoclonal antibodies directed against Fucosyl-GM1. Cancer Res. *46:* 1403 (1986).
500 Nossal, G.J.V.; Szenberg, A.; Ada, G.L.; Austin, C.M.: Single cell studies on 19S antibody production. J. exp. Med. *119:* 485 (1964).
501 Nudelman, E.; Hakomori, S.; Kannagi, R.; Levery, S.; Yeh, M.-Y.; Hellström, K.E.; Hellström, I.: Characterization of a human melanoma-associated ganglioside antigen defined by a monoclonal antibody. J. biol. Chem. *257:* 12752 (1982).
502 Nudelman, E.; Kannagi, R.; Hakomori, S.; Parsons, M.; Lipinski, M.; Wiels, J.; Fellous, M.; Tursz, T.: A glycolipid antigen associated with Burkitt lymphoma defined by a monoclonal antibody. Science (Wash. DC) *220:* 509 (1983).
503 Nuti, M.; Teramoto, Y.A.; Mariani-Constantini, R.; Horan Hand, P.; Colcher, D.; Schlom, J.: A monoclonal antibody (B72.3) defines patterns of distribution of a novel tumor-associated antigen in human mammary carcinoma cell populations. Int. J. Cancer *29:* 539 (1982).
504 Oberhausen, E.; Steinstraesser, A.; Schroth, H.J.; Klein, A.; Berberich, R.: Immunoscintigraphy of colorectal tumors with the MAb 431/31; in Höfer, Bergmann, Radioaktive Isotope in Klinik und Forschung, vol. 17, p. 48 (Egermann, Wien 1986).
505 Oberhausen, E.; Klein, A.; Berberich, R.: Clinical trial of MAb 431/26 and MAb 431/31 labeled with different nuclides. Nucl. Med. *26:* suppl., p. 51 (1987).
506 Oettgen, H.F.; Hellstrom, K.E.: Principles of immunology: Tumor immunology; in Holland, Frei, Cancer medicine, p. 1029 (Lea & Febiger, Philadelphia 1982).
507 Ohoka, H.; Shinomiya, H.; Yokoyama, M.; Ochi, K.; Takeuchi, M.; Utsumi, S.: Thomsen-Friedenreich antigen in bladder tumors as detected by specific antibody: a possible marker of recurrence. Urol. Res. *13:* 47 (1985).
508 Okabe, T.; Kaizu, T.; Fujisowa, M.; Watanabe, J.; Kojima, K.; Yamashita, T.; Takaku, F.: Monoclonal antibodies to surface antigens of small cell carcinoma of the lung. Cancer Res. *44:* 5273 (1984).
509 Okabe, T.; Kaizu, T.; Ozawa, K.; Urabe, A.; Takaku, F.: Elimination of small cell lung cancer cells in vitro from human bone marrow by a monoclonal antibody. Cancer Res. *45:* 1930 (1985).

510 Okada, Y.; Matsuura, H.; Hakomori, S.-I.: Inhibition of tumor cell growth by aggregation of a tumor-associated glycolipid antigen: A close functional association between gangliotriaosylceramide and transferrin receptor in mouse lymphoma L-5178Y. Cancer Res. *45:* 2793 (1985).

511 Oldham, R. K.; Foon, K. A.; Morgan, Ch.; Woodhouse, C. S.; Schroff, R. W.; Abrams, P. G.; Fer, M.; Schoenberger, C. S.; Farrell, M.; Kimball, E.; Sherwin, S. A.: Monoclonal antibody therapy of malignant melanoma: In vivo localization in cutaneous metastasis after intravenous administration. J. clin. Oncol. *2:* 1235 (1984).

512 Olsson, L.; Kronstrom, H.; Cambon-de Mouzon, A.; Honsik, C.; Brodin, T.; Jakobsen, B.: Antibody producing human-human hybridomas. 1. Technical aspects. J. immunol. Methods *61:* 17 (1983).

513 Olsson, L.; Andreasen, R. B.; Ost, A.; Christensen, B.; Bieberfeld, P.: Antibody producing human-human hybridomas. II. Derivation and characterization of an antibody specific for human leukemia cells. J. exp. Med. *159:* 537 (1984).

514 Order, S. E.; Klein, J.; Ettinger, D.; Alderson, P.; Siegelman, S.; Leichner, P.: Phase I–II study of radiolabelled antibody integrated in the treatment of primary hepatic malignancies. Int. J. Radiat. Oncol. Biol. Phys. *6:* 703 (1980).

515 Ortho Multicenter Transplant Study Group: A randomized clinical trial of OKT-3 monoclonal antibody for acute rejection of cadaveric renal transplant. New Engl. J. Med. *313:* 337 (1985).

516 Osieka, R.; Houchens, D. P.; Golding, A.; Johanson, R. K.: Chemotherapy of human colon xenografts in athymic (nude) mice. Cancer *40:* 2640 (1977).

517 Otterness, I.; Krush, F.: Principles of antibody reactions; in Marchalonis, Farr, Antibody as a tool, p. 97 (John Wiley, New York 1982).

518 Paik, C. H.; Ebbert, M. A.; Murphy, P. R.; Lassman, C. R.; Reba, R. C.; Eckelman, W. C.; Pak, K. Y.; Pow, J.; Steplewski, J.; Koprowski, H.: Factors influencing DTPA conjugation with antibodies by cyclic DTPA anhydride. J. nucl. Med. *24:* 1158 (1983).

519 Paik, C. H.; Phan, L. N. B.; Hong, J. J.; Sahami, M. S.; Heald, S. C.; Reba, R. C.; Steigman, J.; Eckelman, W. C.: The labelling of high affinity sites of antibodies with ^{99m}Tc. Int. J. nucl. Med. Biol. *12:* 3 (1985).

520 Pant, K. D.; Ram, M. D.; Sachatello, C. R.; Hagihari, P. F.; Griffin, W. O.; Van Nagell, J. R.: Identification of a new common antigen of malignant human colorectal and mucinous ovarian tumours (COTA). Oncodevelopmental Biology and Medicine, XI Annual Meeting, Stockholm, Sweden, p. 25 (1983).

521 Papsidero, L. D.; Croghan, G. A.; Johnson, E. A.; Chu, T. M.: Immunoaffinity isolation of ductal carcinoma antigen using monoclonal antibody F36/22. Mol. Immunol. *21:* 955 (1984).

522 Pateisky, N.; Philipp, K.; Skodler, W. D.; Czerwenka, K.; Hamilton, G.; Burchell, J.: Radioimmunodetection in patients with suspected ovarian cancer. J. nucl. Med. *26:* 1369 (1985).

523 Paulie, S.; Lundblad, M.-L.; Hansson, Y.; Koho, H.; Ben-Aissa, H.; Perlmann, P.: Production of antibodies to cellular antigens by EBV-transformed lymphocytes from patients with urinary bladder carcinoma. Scand. J. Immunol. *20:* 461 (1984).

524 Perkins, A. C.; Pimm, M. V.; Birch, M. K.: The preparation and characterization of ^{111}In-labelled 791 T/36 monoclonal antibody for tumour immunoscintigraphy. Eur. J. nucl. Med. *10:* 296 (1985).

525 Perkins, A. C.; Armitage, N. C.; Harrison, R. C.; Riley, A. L. M.; Wastie, M. L.;

Hardcastle, J.D.: Planar and SPECT imaging of colon cancer using ^{111}In-Anti-CEA monoclonal antibody (C46); in Schmidt, Ell, Britton, Nuklearmedizin in Forschung und Praxis, p. 472 (Schattauer, Stuttgart 1986).

526 Philben, V.J.; Jakowatz, J.G.; Beatty, B.G.; Vlahos, W.G.; Paxton, R.J.; Williams, L.E.; Shively, J.E.; Beatty, J.D.: The effect of tumor CEA content and tumor size on tissue uptake of Indium 111-labeled anti-CEA monoclonal antibody. Cancer 57: 571 (1986).

527 Philip, T.; Zucker, J.M.; Farrot, M.; Bordigono, A.; Plouvier, A.; Robert, J.L.; Bernard, G.; Soullet, J., Philip, J.P.; Lutz, J.P.; Carton, P.; Kemshead, J.: Purged autologous bone marrow transplantation in 25 cases of very poor prognosis neuroblastoma. Lancet ii: 576 (1985).

528 Philpott, G.W.; Bower, R.J.; Parker, C.W.: Improved selective cytotoxicity with an antibody-diphtheria toxin conjugate. Surgery 73: 928 (1973).

529 Pimm, M.V.; Jones, J.A.; Price, M.R.; Middle, J.G.; Embleton, M.J.; Baldwin, R.W.: Tumour localization of monoclonal antibody against a rat mammary carcinoma and suppression of tumour growth with adriamycin-antibody conjugates. Cancer Immunol. Immunother. 12: 125 (1982).

530 Pimm, M.V.; Rowland, G.F.; Simmonds, R.G.; Marsden, H.; Embleton, M.J.; Jacobs, E.; Baldwin, R.W.: Suppression of growth of a human tumour xenograft by a vindesine monoclonal antibody conjugate. Br. J. Cancer 49: suppl., p. 384 (1984).

531 Pimm, M.V.; Perkins, A.C.; Armitage, N.C.; Baldwin, R.W.: The characteristics of blood-borne radiolabels and the effect of anti-mouse IgG antibodies on localisation of radiolabelled monoclonal antibody in cancer patients. J. nucl. Med. 26: 1011 (1985).

532 Pimm, M.V.; Armitage, N.C.; Perkins, A.C.; Smith, W.; Baldwin, R.W.: Localization of an anti-CEA monoclonal antibody in colorectal carcinoma xenografts. Cancer Immunol. Immunother. 19: 8 (1985).

533 Pimm, M.V.; Perkins, A.C.; Armitage, N.C.; Baldwin, R.W.: Localization of anti-osteogenic sarcoma monoclonal antibody 791T/36 in a primary human osteogenic sarcoma and its subsequent xenograft in immunodeprived mice. Cancer Immunol. Immunother. 19: 18 (1985).

534 Pinto, H.; Lerario, A.C.; Torres de Toledo e Souza, I.; Wajchenberg, B.L.; Mattar, E.; Pieroni, R.R.: Preparation of high-quality Iodine-125-labelled pituitary human follicle-stimulating hormone (hFSH) for radioimmunoassay: Comparison of enzymatic and chloramine-T iodination. Clin. Chim. Acta 76: 25–34 (1977).

535a Pirker, R.; Fitzgerald, D.J.P.; Hamilton, T.C.; Ozols, R.F.; Willingham, M.C.; Pastan, I.: Anti-transferrin receptor antibody linked to Pseudomonas exotoxin as a model immunotoxin in human ovarian carcinoma cell lines. Cancer Res. 45: 751 (1985).

535b Poljak, R.J.; Amzel, L.M.; Avey, H.P.; Chen, B.L.; Phizackerley, R.P.; Saul, F.: Three-dimensional structure of the Fab' fragment of a human immunoglobulin at 2.8-A resolution. Proc. natn. Acad. Sci. USA 70: 3305 (1973).

536 Poncelet, P.; Blythman, H.E.; Carriere, D.; Casellas, P.; Dussossoy, D.; Gros, O.; Gros, P.; Jansen, F.K.; Laurent, J.C.; Liance, M.C.; Vidal, H.; Voisin, G.A.: Present potential of immunotoxins. Behring Inst. Mitt. 74: 94 (1984).

537 Powe, J.; Herlyn, D.; Alavi, A.; et al.: Radioimmunodetection of human tumor xenografts by monoclonal antibodies correlates with antibody density and affinity; in Britton, Donato, Immunoscintigraphy, p. 139 (Gordon and Breach, New York 1985).

538 Prehn, R.T.; Main, J.M.: Immunity to methylcholanthrene-induced sarcoma. J. natn. Cancer Inst. 18: 759 (1957).

539 Pressman, D.; Keighley, G.: The zone of activity of antibodies as determined by the use of radioactive tracers; the zone of an activity of nephrotoxic antikidney serum. J. Immunol. *59:* 141 (1948).
540 Pressman, D.; Korngold, L.: The in vivo localizaton of anti-Wagner-osteogenic-sarcoma antibodies. Cancer *6:* 619 (1953).
541 Pressman, D.: The development and use of radiolabeled antitumor antibodies. Cancer Res. *40:* 2960 (1980).
542 Price, M. R.; Hannant, D.; Embleton, M. J.; Baldwin, R. W.: Icrew workshop report: detection and isolation of tumor-associated antigens. Br. J. Cancer *41:* 843 (1980).
543 Primus, F. J.; Wang, R. H.; Goldenberg, D. M.; Hansen, H. J.: Localization of human GW-39 tumors in hamsters by radiolabeled heterospecific antibody to carcinoembryonic antigen. Cancer Res. *33:* 2977 (1973).
544 Primus, F. J.; Newell, K. D.; Blue, N.; Goldenberg, D. M.: Immunological heterogeneity of carcinoembryonic antigen: antigenic determinants on carcinoembryonic antigen distinguished by monoclonal antibodies. Cancer Res. *43:* 686 (1983).
545 Primus, F. J.; Freeman, J. W.; Goldenberg, D. M.: Immunological heterogeneity of carcinoembryonic antigen: purification from meconium of an antigen related to carcinoembryonic antigen. Cancer Res. *43:* 679 (1983).
546 Primus, F. J.; Kuhns, J. K.; Goldenberg, D. M.: Immunological heterogeneity of carcinoembryonic antigen determinants in colonic tumors with monoclonal antibodies. Cancer Res. *43:* 693 (1983).
547 Pukel, C. S.; Lloyd, K. O.; Trabassos, L. R.; Dippold, W. G.; Oettgen J. F.; Old, L. J.: GD_3, a prominent ganglioside of human melanoma: detection and characterization by mouse monoclonal antibody. J. exp. Med. *155:* 1333 (1982).
548 Quinones, J.; Mizejewski, G.; Beierwaltes, W. H.: Choriocarcinoma scanning using radiolabeled antibody to chorionic gonadotropic. J. nucl. Med. *12:* 69 (1971).
549 Rainsbury, R. M.: The localization of human breast carcinomas by radiolabelled monoclonal antibodies. Br. J. Surg. *71:* 805 (1984).
550 Rainsbury, R. N.; Westwood, J. H.: Tumor localization with monoclonal antibody radioactively labeled with metal chelate rather than iodine. Lancer *ii:* 1347 (1982).
551 Ramsay, N.; LeBien, T.; Nesbit, M.; McGlave, P.; Weisdorf, D.; Kenyon, P.; Hurd, D.; Goldman, A.; Kim, T.; Kersey, J.: Autologous bone marrow transplantation for patients with acute lymphoblastic leukemia in second or subsequent remission: results of bone marrow treated with monoclonal antibodies BA-1, BA-2 and BA-3 plus complement. Blood *667:* 508 (1985).
552 Rankin, E. M.; Hekman, A.; Somers, R.; ten Bokkel Huimink, W.: Treatment of two patients with B-cell lymphoma with monoclonal anti-idiotype antibodies. Blood *65:* 1373 (1985).
553 Raso, V.; Antibody mediated delivery of toxic molecules to antigen bearing target cells. Immunol. Rev. *62:* 93 (1982).
554 Raso, V.; Ritz, J.; Basala, M.; Schlossman, S. F.: Monoclonal antibody-ricin A chain conjugate selectively cytotoxic for cells bearing the common acute lymphocytic leukemia antigen. Cancer Res. *42:* 457 (1982).
555 Raso, V.; Griffin, T.: Specific cytotoxicity of a human immunoglobulin-directed Fab'-ricin A chain conjugate. J. Immunol. *125:* 2610 (1980).
556 Raychaudhuri, S.; Saeki, Y.; Fuji, H.; Kohler, H.: Tumor-specific idiotype vaccines. I. Generation and characterization of internal image tumor antigen. J. Immunol. *137:* 1743 (1986).

557 Reagan, K.J.; Wunner, W.H.; Wiktor, T.J.; Koprowski, H.: Antiidiotypic antibodies induce neutralizing antibodies to rabies virus glycoprotein. J. Virol. 48: 660 (1983).
558 Rector, E.S.; Schwenk, R.J.; Tse, K.S.; Sehon, A.H.: A method for the preparation of protein-protein conjugates of predetermined composition. J. immunol. Methods 24: 321 (1978).
559 Redwood, W.R.; Tom, T.D.; Strand, M.: Specificity, efficacy, and toxicity of radioimmunotherapy in erythroleukemic mice. Cancer Res. 44: 5681 (1984).
560 Reeve, J.G.; Wulfrank, D.A.; Stewart, J.; Twentyman, P.R.; Baillie-Johnson, H.: Monoclonal-antibody-defined human lung tumour cell-surface antigens. Int. J. Cancer 35: 769 (1985).
561 Reiner, R.; Siebeneick, H.U.; Christensen, I.; Doering, H.: Chemical modification of enzymes. III. Cross linked alpha chymotrypsin. J. Mol. Catal. 2: 335 (1977).
562 Reiners, B.; Thranhardt, H.; Herrmann, J.; Fuchs, D.; Prager, J.; Zintl, F.; Plenert, W.: Ex-vivo-Behandlung von Knochenmark zur autologen Reinfusion – Erste Erfahrungen mit einem Cocktail aus drei monoklonalen Antikörpern (VIB-Pool). Z. klin. Med. 42: 405 (1987).
563 Reinherz, E.L.; Schlossman, S.F.: The differentiation and function of human lymphocytes. Cell 19: 821 (1980).
564 Reisfeld, T.A.; Greene, M.I.; Yachi, A.: Monoclonal antibodies – progress in cancer immunobiology and clinical application. Cancer Res. 46: 2193 (1986).
565 Reynolds, C.P.; Seeger, R.C.; Vo, D.D.; Black, A.T.; Wells, J.; Ugelstad, J.: Model system for removing neuroblastoma cells from bone marrow using monoclonal antibodies and magnetic immunobeads. Cancer Res. 46: 5882 (1986).
566 Rhodes, B.A.; Zamora, P.O.; Newell, K.D.; Valdez, E.F.: Technetium-99m-labeling of murine monoclonal antibody fragments. J. nucl. Med. 27: 685 (1986).
567 Richardson, A.P.; Mountford, P.J.; Baird, A.C.; Heyderman, E.; Richardson, T.C.; Coackley, A.J.: An improved iodogen method of labelling antibodies with I-123. Nucl. Med. Commun. 7: 355 (1986).
568 Riechmann, L., Winter, G., MRC Cambridge, UK, personal communication (1987).
569 Ritz, J.; Pesando, J.M.; Sallan, S.E.; Clavell, L.A.; Notis-McConathy, J.; Rosenthal, P.; Schlossman, S.F.: Serotherapy of acute lymphoblastic leukemia with monoclonal antibody. Blood 58: 141 (1981).
570 Ritz, J.; Pesando, J.M.; Notis-McConarty, J.; Schlossman, S.F.: Modulation of human acute lymphoblastic leukemia antigen induced by monoclonal antibody in vitro. J. Immunol. 125: 1506 (1980).
571 Ritz, J.; Schlossman, S.F.: Utilization of monoclonal antibodies in the treatment of leukemia and lymphoma. Blood 59: 1 (1982).
572 Ritz, J.; Bast, R.C., Jr.; Clavell, L.; Hercend, T.; Sallan, S.E.; Lipton, J.M.; Fenney, M.; Nathan, D.G.; Schlossman, S.F.: Autologous bone marrow transplantation in CALLA positive acute lymphoblastic leukemia after in vitro treatment with J5 monoclonal antibody and complement. Lancet ii: 60 (1982).
573 Ritz, J.; Pesando, J.M.; Sallan, S.E.; Clavell, L.A.; Notis-McConarty, J.; Rosenthal, P.; Schlossman, S.F.: Serotherapy of acute lymphoblastic leukemia with monoclonal antibody. Blood 58: 141 (1985).
574 Riva, P.; Paganelli, G.; Tison, V.; et al.: Melanoma immunoscintigraphy with ^{99}Tc-labelled monoclonal F(ab')$_2$: Sensitivity and specificity studies; in Donato, Britton, Immunoscintigraphy p. 217 (Gordon and Breach, New York 1985).

575 Riva, P.; Paganelli, G.; Moscatelli, G.; Benini, S.: Therapeutic use of radiolabelled monoclonal antibodies by systemic and locoregional administration. Proceedings 'Advances in the applications of monoclonal antibodies in clinical oncology', p. 30, London (1987).
576 Roberts, S.; Cheetham, J. C.; Rees, A. R.: Generation of an antibody with enhanced affinity and specificity for its antigen by protein engineering. Nature, Lond. *328:* 731 (1987).
577 Rodeck, U.; Herlyn, M.; Herlyn, D.; Molthoff, C.; Atkinson, B.; Varello, M.; Steplewski, Z.; Koprowski, H.: Tumor growth modulation by a monoclonal antibody to the epidermal growth factor receptor: immunologically mediated and effector cell-dependent effects. Cancer Res. *47:* 3692 (1987).
578 Rösler, H.; Noelpp. U.: Zur Dosierung von therapeutischen Alpha-Strahlern. 3. Böttsteiner Kolloquium Radioimmunszintigraphie, EIR, p. 70, Würenlingen (1985).
579 Rogers, G. T.; Pedley, R. B.; Boden, J.; Harwood, P. J.; Bagshawe, K. D.: Effect of dose escalation of a monoclonal anti-CEA IgG on tumour localisation and tissue distribution in nude mice xenografted with human colon carcinoma. Cancer Immunol. Immunother. *23:* 107 (1986).
580 Rogers, G. T.; Harwood, P. J.; Pedley, R. B.; Boden, J.; Bagshawe, K. D.: Dose-dependent localisation and potential for therapy of F(ab')$_2$ fragments against CEA studied in a human tumour xenograft model. Br. J. Cancer (in press).
581 Rosen, S. T.; Zimmer, A. M.; Goldman-Leikin, R.; Gordon, L. I.; Kazikiewiez, J. M.; Kaplan, E. H.; Variakojis, D.; Marder, R. J.; Dykewicz, M. S.; Pergies, A.: Radioimmunodetection and radioimmunotherapy of cutaneous T-cell lymphomas using an ^{131}I-labelled monoclonal antibody: An Illinois Center Council Study. J. clin. Oncol. *5:* 562 (1987).
582 Rosenberg, R. A.; Murray, T. M.: The mechanisms of methionine oxidation concomitant with hormone radioiodination: Comparative studies of various oxidants using a simple new method. Biochim. biophys. Acta *584:* 261 (1979).
583 Ross, W. C. J.: The conjugation of chlorambucil with human gamma-globulin. Chem. Biol. Interactions *10:* 169 (1975).
584 Ross, A. H.; Herlyn, D.; Iliopoulos, D.; Koprowski, H.: Isolation and characterization of a carcinoma associated antigen. Biochem. biophys. Res. Commun. *135:* 297 (1986).
585 Rowland, G. F.; O'Neill, G. J.; Davies, D. A. L.: Suppression of tumour growth in mice by a drug-antibody conjugate using a novel approach to linkage. Nature, Lond. *255:* 487 (1975).
586 Rowland, G. F.: Effective antitumour conjugates of alkylation drug and antibody using dextran as the intermediate carrier. Eur. J. Cancer *13:* 593 (1977).
587 Rowland, G. F.; Simmonds, R. G.; Corvalan, J. R. F.; Marsden, C. H.; Johnson, J. R.; Woodhouse, C. S.; Ford, C. H. J.; Newman, C. E.: The potential use of monoclonal antibodies in drug targeting. Protides biol. Fluids *29:* 921 (1981).
588 Rowland, G. F.; Axton, C. A.; Baldwin, R. W.; Brown, J. P.; Corvalan, J. R. F.; Embleton, M. J.; Gore, V. A.; Hellström, I.; Hellström, K. E.; Jacobs, E.; Marsden, C. H.; Pimm, M. V.; Simmonds, R. G.; Smith, W.: Antitumor properties of vindesine-monoclonal antibody conjugates. Cancer Immunol. Immunother. *19:* 1 (1985).
589 Rowland, G. F.; Simmonds, R. G.; Gore, V. A.; Marsden, C. H.; Smith, W.: Drug localisation and growth inhibition studies of vindesine-monoclonal anti-CEA conjugates in a human tumour xenograft. Cancer Immunol. Immunother. *21:* 183 (1986).

590 Rozman, C.; Vdilella, R.; Vives, J.: Viral particles in hybridoma clones secreting monoclonal antibodies. Lancet ii: 445 (1982).
591 Sacks, D.L.; Esser, K.M.; Sher, A.: Immunization of mice against African trypanosomiasis using anti-idiotypic antibodies. J. exp. Med. 155: 1108 (1982).
592 Sahagan, B.G.; Dorai, H.; Saltzgaber-Muller, J.; Toneguzzo, F.; Guindon, C.A.; Lilly, S.P.; McDonald, K.W.; Morrissey, D.V.; Stone, B.A.; Davis, G.L.; McIntosh, P.K.; Moore, G.P.: A genetically engineered murine/human chimeric antibody retains specificity for human tumor-associated antigen. J. Immunol. 137: 1066 (1986).
593 Saito, M.; Yu, R.K.; Cheung, N.-K.V.: Ganglioside GD2 specificity of monoclonal antibodies to human neuroblastoma cells. Biochem. biophys. Res. Commun. 127: 1 (1985).
594 Saji, S.; Zylstra, S.; Schepart, B.S.; Ghosh, S.K.; Jou, Y.-H.; Takita, H.; Bankert, R.: Monoclonal antibodies specific for two different histological types of human lung carcinoma. Hybridoma 3: 119 (1983).
595 Scatchard, G.: The attraction of proteins for small molecules and ions. Ann. N.Y. Acad. Sci. 51: 660 (1949).
596 Searle, F.; Bagshawe, K.D.; Begent, R.H.J.; Jewkes, R.F.; Jones, B.E.; Keep, P.A.; Lewis, J.C.M.; Vernon, P.: Radioimmunolocalisation of tumours by external scintigraphy after administration of sup 1sup 3sup 1I antibody to carcinoembryonic antigen. Nucl. med. Comm. 1: 131 (1980).
597 Sears, H.F.; Atkinson, B.; Mattis, J.; Ernst, C.; Herlyn, D.; Steplewski, Z.; Haeyry, P.; Koprowski, H.: Phase-1 clinical trials of monoclonal antibody in treatment of gastrointestinal tumours. Lancet I: 762 (1982).
598 Sears, H.F.; Herlyn, D.; Steplewski, Z.; Koprowski, H.: Effects of monoclonal antibody immunotherapy on patients with gastrointestinal adenocarcinoma. J. Biol. Response Modif. 3: 138 (1984).
599 Sears, H.F.; Herlyn, D.; Steplewski, Z.; Koprowski, H.: Phase II clinical trial of a murine monoclonal antibody cytotoxic for gastrointestinal adenocarcinoma. Cancer Res. 45: 5910 (1985).
600 Sedlacek, H.H.: Pathophysiological aspects of immune complex diseases. Part I. Interaction with plasma enzyme systems, cell membranes and the immune response. Klin. Wschr. 58: 543 (1980).
601 Sedlacek, H.H.: Pathophysiological aspects of immune complex diseases. Part II. Phagocytosis, exocytosis, and pathogenic depositions. Klin. Wschr. 58: 593 (1980).
602 Sedlacek, H.H.; Gronski, P.; Hofstaetter, T.; Kanzy, E.J.; Schorlemmer, H.U.; Seiler, F.R.: The biological properties of immunoglobulin G and its split products ($F(ab')_2$ and Fab). Klin. Wschr. 61: 723 (1983).
603 Sedlacek, H.H.; Dickneite, G.; Schorlemmer, H.U.: Chemotherapeutics: A questionable or a promising project. Comp. Immunol. Microbiol. Infect. Dis. 9: 99 (1986).
604 Sedlacek, H.H.: Tumorimmunologie und Tumortherapie – Eine Standortbestimmung. Contr. Oncol., vol. 25 (Karger Basel 1986).
605a Seeger, R.; Lenarsky, C.; Moss, T.; Feig, S.; Selch, M.; Ramasay, N.; Harris, R.; Reynolds, C.P.; Siegel, S.; Sather, H.; Hammond, D.; Wells, J.: Bone marrow transplantation (BMT) for poor prognosis neuroblastoma. Proc. American Soc. Clin. Oncol. 6: 221 (1987).
605b Seiler, F.R.; Gronski, P.; Kurrle, R.; Lüben, G.; Harthus, H.-P.; Ax, W.; Bosslet, K.; Schwick, H.-G.: Monoklonale Antikörper: Chemie, Funktion und Anwendungsmöglichkeiten. Angew. Chem. 97: 141 (1985).

606 Sell, S.; Reisfeld, R. A. (eds.): Monoclonal antibodies in cancer (Humana Press, Clifton, N. J. 1985).
607 Senekowitsch, R.; Glaessner, H.; Reidel, G.; Möllenstaedt, S.; Kriegel, H.; Pabst, H. W.: Experimentelle Studien zur Radioimmuntherapie. Nucl. Med. *26:* suppl., p. 45 (1987).
608 Seto, M.; Umemoto, N.; Saito, M.; Masuho, Y.; Hara, T.; Takahashi, T.: Monoclonal anti-MM46 antibody: ricin A chain conjugate: in vitro and in vivo antitumor activity. Cancer Res. *42:* 5209 (1982).
609 Shah, S. A.; Gallagher, B. M.; Sands, H.: Radioimmunodetection of small human tumor xenografts in spleen of athymic mice by monoclonal antibodies. Cancer Res. *45:* 5824 (1985).
610 Sharkey, R. M.; Primus, F. J.; Goldenberg, D. M.: Second antibody clearance of radiolabeled antibody in cancer radioimmunodetection. Proc. natn. Acad. Sci. USA *81:* 2843 (1984).
611 Shawler, D. L.; Bartholomew, R. M.; Smith, L. M.; Dillman, R. O.: Human immune response to multiple injections of murine monoclonal IgG. J. Immunol. *135:* 1530 (1985).
612 (See reference no. 611).
613 Shen, W.-C.; Ryser, H. J.-P.: Cis-aconityl spacer between daunomycin and macromolecular carriers: a model of Ph-sensitive linkage releasing drug from a lysosomotropic conjugate. Biochem. biophys. Res. Commun. *102:* 1048 (1981).
614 Shen, W.-C.; Ryser, H. J.-P.: Selective killing of Fc-receptor-bearing tumor cells through endocytosis of a drug-carrying immune complex. Proc. natn. Acad. Sci. USA *81:* 1445 (1984).
615 Shiku, H.; Takahashi, T.; Oettgen, H. F.; Old, L. J.: Cell surface antigens of human malignant melanoma. II. Serological typing with immune adherence assays and definition of two new surface antigens. J. exp. Med. *144:* 873 (1976).
616 Siddiqui, B.; Whitehead, J. S.; Kim, Y. S.: Glycosphingolipids in human colonic adenocarcinoma. J. biol. Chem. *253:* 2168 (1978).
617 Siddiqui, B.; Buehler, J.; De Gregorio, M. W.; Macher, B.: Differential expression of ganglioside GD_3 by human leukocytes and leukemia cells. Cancer Res. *44:* 5262 (1984).
618 Sikora, K.; Alderton, T.; Phillips, J.; Watson, J. V.: Human hybridomas from malignant gliomas. Lancet *ii:* 11 (1982).
619 Sikora, K.; Alderson, T.; Ellis, J.; Phillips, J.; Watson, J.: Human hybridomas from patients with malignant disease. Br. J. Cancer *47:* 135 (1983).
620 Sikora, K.: Human monoclonal antibodies to cancer cells; in Strelkauskas, Immunology Series, vol. 13, Human hybridomas; Diagnostic and therapeutic applications, p. 159, (Marcel Dekker, New York 1987).
621 Sindelar, W. F.; Mahrer, M. M.; Herlyn, D.; Sears, H. F.; Steplewski, Z.; Koprowski, H.: Trial of therapy with monoclonal antibody 17-1A in pancreatic carcinoma: Preliminary results. Hybridoma *5:* suppl., p. 125 (1986).
622 Sklar, J.; Cleary, M. L.; Thielemans, K.; Gralow, J.; Warnek, R.; Levy, R.: Biclonal B-cell lymphoma. New Engl. J. Med. *311:* 20 (1984).
623 Smedley, H. M.; Finan, P.; Lennox, E. S.; Ritson, A.; Takei, F.; Wraight, P.; Sikora, K.: Localisation of metastatic carcinoma by a radiolabelled monoclonal antibody. Br. J. Cancer *47:* 253 (1983).
624 Smyth, M. J.; Pietersz, G. A.; McKenzie, I. F. C.: Selective enhancement of antitumor activity of N-acetyl melphalan upon conjugation to monoclonal antibodies. Cancer Res. *47:* 62 (1987).

625 Smyth, M.J.; Pietersz, G.A.; McKenzie, I.F.C.: The in vitro and in vivo anti-tumour activity of N-AcMEL-(Fab')$_2$ conjugates. Br. J. Cancer 55: 7 (1987).
626 Snyder, H.H., Jr.; Hardy, W.D., Jr.; Zuckerman, E.E.; Fleissner, E.: Characterisation of a tumour-specific antigen on the surface of feline lymphosarcoma cells. Nature, Lond. 275: 656 (1978).
627 Sordat, B.; Wang, W.R.: Human colorectal tumor xenografts in nude mice: expression of malignancy. Behring Inst. Mitt. 74: 291 (1984).
628 Spitler, L.E.: Immunotoxin therapy of malignant melanoma. Med. Oncol. & Tumor Pharmacother. 3: 147 (1986).
629 Spitler, L.E.; Rio, M. del; Khentigan, A.; Wedel, N.I.; Brophy, N.A.; Miller, L.L.; Harkonen, W.S.; Rosendorf, L.L.; Lee, H.M.; Mischak, R.P.; Kawahata, R.T.; Stoudemire, J.B.; Fradkin, L.B.; Bautista, E.E.; Scannon, P.J.: Therapy of patients with malignant melanoma using a monoclonal antimelanoma antibody-ricin A chain immunotoxin. Cancer Res. 47: 1717 (1987).
630 Springer, G.F.; Desai, P.R.; Fry, W.A.; Goodale, R.L.; Shearen, J.G.; Scanlon, E.F.: T antigen, a tumor marker against which breast, lung, pancreas carcinoma patients mount immune responses. Cancer Detect. Prevent. 6: 111 (1983).
631 Srivastava, P.C.; Knapp, F.F.; Dickson, D.R.; Alred, J.F.: Design and synthesis of a new N-(p-Iodophenyl)-maleimide-kit for labeling of antibodies with I–131 and I–123. J. nucl. Med. 28: 726 (1987).
632 Sugita, K.; Majdic, O.; Stockinger, H.; Holter, W.; Köller, U.; Peschel, C.; Knapp, W.: Use of a cocktail of monoclonal antibodies and human complement in selective killing of acute lymphocytic leukemic cells. Int. J. Cancer 37: 351 (1986).
633 Sumner, W.C.; Foraker, A.G.: Spontaneous regression of human melanoma. Clinical and experimental studies. Cancer 13: 179 (1960).
634 Sun, L.K.; Curtis, P.; Rakowicz-Szulczynska, E.; Ghrayeb, J.; Chang, N.; Morrison, S.L.; Koprowski, H.: Chimeric antibody with human constant regions and mouse variable regions directed against carcinoma-associated antigen 17–1A. Proc. natn. Acad. Sci. USA 84: 214 (1987).
635 Svenberg, T.; Hammarström, S.; Hedin, A.: Purification and properties of biliary glycoprotein (BGP I). Immunochemical relationship to carcinoembryonic antigen. Mol. Immunol. 16: 245 (1979).
636 Scharff, M.D.; Roberts, S.; Thammana, P.: Hybridomas as a source of antibodies. Hosp. Pract. 16: 61 (1981).
637 Scheidhauer, K.; Dennecke, H.; Moser, E.: Radioimmunoscintigraphy using ECT in comparison to TCT in the follow-up of colorectal cancer. Nucl. Med. 25: A51 (1986).
638 Scheidhauer, K.; Stefani, F.; Markl, A.; Schuhmacher, U.: Wehmeyer, G.; Moser, E.: Radioimmunoscintigraphy with ^{99m}Tc-labeled monoclonal antibodies in primary ocular melanoma. Nucl. Med. 25: A51 (1986).
639 Scheinberg, D.A.; Strand, M.; Gansow, O.A.: Tumor imaging with radioactive metal chelates conjugated to monoclonal antibodies. Science 215: 1511 (1982).
640 Scheinberg, D.A.; Strand, M.: Leukemic cell targeting and therapy by monoclonal antibody. Cancer Res. 42: 44 (1982).
641 Scheinberg, D.A.; Strand, M.: Kinetic and catabolic considerations of monoclonal antibody targeting in erythroleukemic mice. Cancer Res. 43: 265 (1983).
642 Schengrund, C.-L.; Repman, M.A.; Shochat, S.J.: Ganglioside composition of human

neuroblastomas: Correlation with prognosis. A Pediatric Oncology Group Study. Cancer 56: 2640 (1985).
643 Schenk, H. von, Larsson, I.; Thorell, J.I.: Improved radioiodination of glucagon with the lactoperoxidase method. Influence of pH on iodine substitution. Clin. Chim. Acta. 69: 225 (1976).
644 Schlom, J.; Greiner, J.; Horan Hand, P.; Colcher, D.; Inghirami, G.; Weeks, M.; Pestka, S.; Fisher, P.B.; Noguchi, P.; Kufe, D.: Monoclonal antibodies to breast cancer-associated antigens as potential reagents in the management of breast cancer. Cancer 54: suppl. 11, p. 2777 (1984).
645 Schlom, J.; Weeks, M.O.: Potential clinical utility of monoclonal antibodies in the management of human carcinomas; in DeVita, Hellman, S. Rosenberg, Important Advances in Oncology, pp. 170 J.B. Lippincott, Philadelphia (1985).
646 Schlom, J.: Basic principles and applications of monoclonal antibodies in the management of carcinomas. The Richard and Hind Rosenthal Foundation Award Lecture. Cancer Res. 46: 3225 (1986).
647 Schmiegel, W.H.; Kalthoff, H.; Arndt, R.; Gieseking, J.; Greten, H.; Klöppel, G.; Kreiker, C.; Ladak, A.; Lampe, V.; Ulrich, S.: Monoclonal antibody-defined human pancreatic cancer-associated antigens. Cancer Res. 45: 1402 (1985).
648 Schorlemmer, H.U.; Bosslet, K.; Kern, H.F.; Sedlacek, H.H.: A monoclonal antibody with binding and inhibiting activity on human pancreatic carcinoma cells. II. Inhibition of functions of pancreatic tumor cells by specific binding with a murine monoclonal antibody. Behring Inst. Mitt. Nr. 82 (in press, 1988).
649 Schorlemmer, H.U.; Bosslet, K.; Kern, H.F.; Sedlacek, H.H.: Inhibition of functions related to human mononuclear phagocytes and human pancreatic adenocarcinoma cells by murine monoclonal antibodies reacting with common surface antigens. Cancer Immunol. Immunother. Behring Inst. Mitt. Nr. 82 (in press, 1988).
650 Schroer, K.R.; Briles, D.E.; Boxel, J.A., van; Davie, J.M.: Idiotypic uniformity of cell surface immunoglobulin in chronic lymphocytic leukemia: evidence for monoclonal proliferation. J. exp. Med. 140: 1416 (1974).
651 Schroff, R.W.; Foon, K.A.; Beatty, S.M.; Oldham, R.K.; Morgan, A.C., Jr.: Human anti-murine immunoglobulin response in patients receiving monoclonal antibody therapy. Cancer Res. 45: 879 (1985).
652 Schuepbach, J.; Kalyanaraman, V.S.; Sarngadharan, M.G.; Blattner, W.A.; Gallo, R.C.: Antibodies against three purified proteins of the human type C retrovirus, human T-cell leukemia-lymphoma virus, in adult T-cell leukemia-lymphoma patients and healthy blacks from the Caribbean. Cancer Res. 43: 886 (1983).
653 Schuhmacher, J.; Matys, R.; Hauser, H.; Maier-Borst, W.; Matzku, S.: Labeling of monoclonal antibodies with a ^{67}Ga phenolic aminocarboxylic acid chelate. Part I: Chemistry and labeling technique. Eur. J. nucl. Med. 12: 397 (1986).
654 Schulz, G.; Bumol, T.F.; Reisfeld, R.A.: Monoclonal antibody directed effector cells selectively lyse human melanoma cells in vitro and in vivo. Proc. natn. Acad. Sci. USA 80: 5407 (1983).
655 Schulz, G.; Cheresh, D.A.; Varki, N.M.; Yu, A.; Staffileno, L.K.; Reisfeld, R.A.: Detection of ganglioside G_{D2} in tumor tissues and sera of neuroblastoma patients. Cancer Res. 44: 5914 (1984).
656 Schulz, G.; Staffileno, L.K.; Reisfeld, R.A.; Dennert, G.: Eradication of established human melanoma tumors by antibody-directed effector cells. J. exp. Med. 161: 1315 (1985).

657 Schulz, G.; Büchler, M.; Muhrer, K. H.; Klapdor, R.; Kübel, R.; Harthus, H. P.; Madry, N.; Bosslet, K.: Immunotherapy of pancreatic cancer with monoclonal antibody BW 494. Int. J. Cancer (in press).
658 Schur, P. H.: IgG subclasses – a review. Ann. Allergy, 58: 89 (1987).
659 Schwaber, J. F.; Posner, M. R.; Schlossman, S. F.; Lazarus, H.: Human-human hybrids secreting pneumococcal antibodies. Hum. Immunol. 9: 137 (1984).
660 Schwarz, A.; Steinstraesser, A.: A novel approach to Tc-99m-labelled monoclonal antibodies. J. nucl. Med. 28: 721 (1987).
661 Schwartz, H. S.: Biochemical action and selectivity of intercalating drugs. Adv. Cancer Chemother. 1: 1 (1979).
662 Stahel, R. A.; Mabry, M.; Sabbath, K.; Speak, J. A.; Bernal, S. D.: Selective cytotoxicity of murine monoclonal antibody LAM 2 against human small-cell carcinoma in the presence of human complement: Possible use for in vitro elimination of tumor cells from bone marrow. Int. J. Cancer 35: 387 (1985).
663 Stanley, C. J.; Paris, F.; Plumb, A.; Webb, A.; Johannsson, A.: Enzyme amplification: A new technique for enhancing the speed and sensitivity of enzyme immunoassays. ICRP 3: 44 (1985).
664 Starling, J. J.; Wright, G. L., Jr.: Disulfide bonding of a human prostate tumor-associated membrane antigen recognized by monoclonal antibody D 83.21. Cancer Res. 45: 804 (1985).
665 Stefanini, M.: Enzymes, isozymes, and enzyme variants in the diagnosis of cancer: a short review. Cancer 55: 1931 (1985).
666 Stein, K. E.; Soderstrom, T.: Neonatal administration of idiotype and anti-idiotype primes for protection against Escherichia coli K13 infection in mice. J. exp. Med. 160: 1001 (1984).
667 Steinstraesser, A.; Schwarz, A.; Kuhlmann, L.; Bosslet, K.: Kinetics of monoclonal antibodies – Influence of labelling (Abstract). J. nucl. Med. 28: 693 (1987).
668 Stepan, D. E.; Bartholomew, R. M.; LeBien, T. W.: In vitro cytodestruction of human leukemic cells using murine monoclonal antibodies and human complement. Blood 63: 1120 (1984).
669 Stephenson, J. R.; Essex, M.; Hino, S.; Mardy, W. D., Jr.; Aaronson, S. A.: Feline oncornavirus-associated cell-membrane antigen (FOCMA): distinction between FOCMA and the major virion glycoprotein. Proc. natn. Acad. Sci. USA 74: 1219 (1977).
670 Steplewski, Z.; Chang, T. H.; Herlyn, M.; Koprowski, H.: Release of monoclonal antibody-defined antigens by human colorectal carcinoma and melanoma cells. Cancer Res. 41: 2723 (1981).
671 Steplewski, Z.; Koprowski, H.: Monoclonal antibody development in the study of colorectal carcinoma-associated antigen. Methods Cancer Res. 20: 285 (1982).
672 Steplewski, Z.; Lubeck, M. D.; Koprowski, H.: Human macrophages armed with murine immunoglobulin G2a antibodies to tumors destroy human cancer cells. Science 221: 865 (1983).
673 Steplewski, Z.; Spira, G.; Blaszczyk, M.; Lubeck, M. D.; Radbruch, A.; Illges, H.; Herlyn, D.; Rajewsky, K.; Scharff, M.: Isolation and characterization of anti-monosialoganglioside monoclonal antibody 19–9 class-switch variants. Proc. natn. Acad. Sci. USA 82: 8653 (1985).
674 Stockinger, H.; Majdic, O.; Liszka, K.; Aberer, W.; Bettelheim, P.; Lutz, D.; Knapp, W.:

Exposure by desialylation of myeloid antigens on acute lymphoblastic leukemia cells. J. natn. Cancer Inst. *73:* 7 (1984).

675 Stong, R. C.: Youle, R. D.; Vallera, D. A.: Elimination of clonogenic T-leukemia cells from human bone marrow using anti Mr 65,000 protein immunotoxins. Cancer Res. *44:* 3000 (1984).

676 Stong, R. C.; Uckin, F.; Youle, R. J.; Kersey, J. H.; Vallera, D. A.: Use of multiple T cell-directed intact ricin immunotoxins for autologous bone marrow transplantation. Blood *66:* 627 (1985).

677 Stramignoni, D.; Bowen, R.; Atkinson, B. F.; Schlom, J.: Differential reactivity of monoclonal antibodies with human colon adenocarcinomas and adenomas. Int. J. Cancer *31:* 543 (1983).

678 Strelkauskas, A. J.; Taylor, C. L.; Aldenderfer, P. H.; Warner, G. A.: Construction of stable human hybrid clones producing antibody reactive with human mammary carcinoma; in Strelkauskas, Immunology series, vol. 13, Human hybridomas, diagnostic and therapeutic applications, p. 227 (Marcel Dekker, New York 1987).

679 Stuart, F. P.; Perdrizet, G.; Steranka, B.; Buckingham, M.; Lopes, D.; Alvarez, V.; Rodwell, J.; McKearn, T.; Hines, J.; Atcher, R.; Friedman, A. M.: Specific destruction of allogeneic tumor cells in vivo by a Bismuth-212 immunotoxin. Transplant. Proc. *19:* 605 (1987).

680 Taetle, R.; Rosen, F.; Abramson, I.; Venditti, J.; Howell, S.: Use of nude mouse xenografts as preclinical drug screens: in vivo activity of established chemotherapeutic agents against melanoma and ovarian carcinoma xenografts. Cancer Treat. Rep. *71:* 297 (1987).

681 Tai, J.; Blair, A. H.; Ghose, T.: Tumor inhibition by chlorambucil covalently linked to antitumor globulin. Eur. J. Cancer *15:* 1357 (1979).

682 Talman, G. L.; Hadjian, R. J.; Morelock, M. M.; Jones, P. L.; Neacy, W.; Liberatore, F. A.; Sands, H.; Gallagher, B. M. Ü.: In vivo tumor localization of technetium-labeled metalothionein/monoclonal antibody conjugates. J. nucl. Med. *25:* 24 (1984).

683 Taniguchi, N.; Yokosawa, N.; Narita, M.; Mitsuyama, T.; Makita, A.: Expression of Forssman antigen synthesis and degradation in human lung cancer. J. natn. Cancer Inst. *67:* 577 1981).

684 Taylor-Papadimitriou, J.; Peterson, J. A.; Arklie, J.; Burchell, J.; Ceriani, R. C.; Bodner, W. F.: Monoclonal antibodies to epithelium specific components of the milk fat globule membrane: production and reaction with cells in culture. Int. J. Cancer *28:* 17–24 (1981).

685 Tempero, M. A.; Pour, P. M.; Uchida, E.; Herlyn, D.; Steplewski, Z.: Monoclonal antibody CO17-1A and leukopheresis in immunotherapy of pancreatic cancer. Hybridoma suppl. 1, p. *5:* 155 (1986).

686 Teng, N. N. H.; Kaplan, H. S.; Herbert, J. M.; Moore, C.; Douglas, H.; Wunderlich, A.; Braude, A. J.: Protection against gram negative bacteremia and endotoxinemia with human monoclonal IgM antibodies. Proc. natn. Acad. Sci. USA *82:* 1790 (1985).

687 Thomas, L.: Discussion; in Von Lawrence, Cellular and humoral aspects of the hypersensitive state, p. 529 (Hoeber, New York 1959).

688 Thompson, C. H.; Stacker, S. A.; Salehi, N.; McKenzie, I. F. C.; Lichtenstein, M.; Leyden, M. J.; Andrews, J. T.: Immunoscintigraphy for detection of lymph node metastases from breast cancer. Lancet *ii:* 1245 (1984).

689 Thompson, C. H.; Lichtenstein, M.; Stacker, S. A.; Leyden, M. J.; Salehi, N.; Andrews,

J. T.; McKenzie, I. F.: Immunoscintigraphy for detection of lymph node metastases from breast cancer. Lancet *ii:* 1245 (1984).
690 Thompson, K. M.; Melamed, M. D.; Eagle, K.; Gorick, B. D.; Gibson, T.; Holburg, A. M.; Hughes-Jones, N. C.: Production of human monoclonal IgG and IgM antibodies with anti-D (Rhesus) specificity using heterohybridomas. Immunology *58:* 157 (1986).
691 Thompson, R. J.; Jackson, A. Pl.; Langlois, N.: Circulating antibodies to mouse monoclonal immunoglobulins in normal subjects – incidence, species specificity, and effects on a two-site assay for creatine kinase-MB isoenzyme. Clin. Chem. *32:* 476 (1986).
692 Thor, A.; Gostein, F.; Ohuchi, N.; Szpak, C. A.; Johnston, W. W.; Schlom, J.: Tumor-associated glycoprotein (TAG-72) in ovarian carcinomas defined by monoclonal antibody B72.3. J. natn. Cancer. Inst. *76:* 995 (1986).
693 Thorpe, P.; Ross, W.; Cumber, A.; Hinson, C.; Edwards, D.; Davies, A.: Toxicity of diphtheria toxin for lymphoblastoid cells is increased by conjugation to anti-lymphocyte globulin. Nature *271:* 752 (1978).
694 Thorpe, P. E.; Brown, A. N.; Ross, W. C.; Cumber, A. J.; Detre, S. I.; Edwards, D. C.; Davies, A. J.; Stirpe, F.: Cytotoxicity acquired by conjugation of an anti-Thy 1.1 monoclonal antibody and the ribosome-inactivating protein, gelonin. Eur. J. Biochem. *116:* 447 (1981).
695 Thorpe, P. E.; Mason, D. W.; Brown, A. N. F.; Simmonds, J.; Ross, W. C. J.; Cumber, A. J.; Forrester, J. A.: Selective killing of malignant cells in a leukaemic rat bone marrow using an antibody-ricin conjugate. Nature, Lond. *297:* 594 (1982).
696 Thorpe, P.; Ross, W.: The preparation and cytotoxic properties of antibody-toxin conjugates. Immunol. Rev. *62:* 119 (1982).
697 Thornton, D.; Nicholas, R. A.: Oncoviruses and monoclonal antibody production. Vet. Rec. *111:* 329 (1982).
698 Treleavan, J. G.; Gibson, F. M.; Ugelstad, J.; Rembau, A.; Phillips, T.; Caine, G. D.; Kemshead, J. T.: Removal of neuroblastoma cells from bone marrow with monoclonal antibodies conjugated to magnetic microspheres. Lancet *i:* 70 (1984).
699 Tritton, T. R.; Yee, G.; Wingard, L. B., Jr.: Immobilized adriamycin: a tool for separating cell surface from intracellular mechanisms. Fed. Proc. *42:* 284 (1983).
700 Trouet, A.; Masquelier, M.; Baurain, R.; Deprez de Campeneere, D.: A covalent linkage between daunorubicin and proteins that is stable in serum and reversible by lysosomal hydrolases, as required for lysosomotropic drug-carrier conjugate. Proc. natn. Acad. Sci. USA *79:* 626 (1982).
701 Trowbridge, I. S.; Domingo, D. L.: Anti-transferrin receptor monoclonal antibody and toxin-antibody conjugates affect growth of human tumour cells. Nature *294:* 171 (1981).
702 Tsuchida, T.; Ravindranath, M. H.; Saxton, R. E.; Irie, R. F.: Gangliosides of human melanoma: altered expression in vivo and in vitro. Cancer Res. *47:* 1278 (1987).
703 Tsukada, Y.; Hurwitz, E.; Kashi, R.; Sela, M.; Hibi, N.; Hara, A.; Hirai, H.: Chemotherapy by intravenous administration of conjugates of daunomycin with monoclonal and conventional anti-rat α-fetoprotein antibodies. Proc. natn. Acad. Sci. USA *79:* 7896 (1982).
704 Tsukada, Y.; Kato, Y.; Umemoto, N.; Takeda, Y.; Hara, T.; Hirai, H.: An anti-alpha-fetoprotein antibody-daunomycin conjugate with a novel poly-L-glutamic acid derivative as intermediate carrier. J. natn. Cancer Inst. *73:* 721 (1984).
705 Tung, E.; Goust, J. M.; Chen, W. Y.; Kang, S. S.; Wang, I. Y.; Wang, A. C.: Cytotoxic effect of anti-idiotype antibody-chlorambucil conjugates against human lymphoblastoid cells. Immunology *50:* 57 (1983).

706 Uadia, P.; Blair, A.H.; Ghose, T.; Ferrone, S.: Uptake of methotrexate linked to polyclonal and monoclonal antimelanoma antibodies by a human melanoma cell line. J. natn. Cancer Inst. *74:* 29 (1985).

707 Uckun, F.M.; Ramakrishnan, S.; Houston, L.L.: Ex vivo elimination of neoplastic cells from human marrow using an anti-Mr 41,000 protein immunotoxin: potentiation by ASTA Z7557. Blut *50:* 19 (1985).

708 Uckun, F.M.; Ramaskrishnan, S.; Haag, D.; Houston, L.L.: Heterogeneity in leukemia cell populations: a clear rationale for use of combination protocols for ex vivo marrow purging. Transplant. Proc. *17:* 462 (1985).

709 Uckun, F.; Ramakrishnan, S.; Houston, L.: Increased efficiency in selective elimination of leukemia cells by a combination of a stable derivative of cyclophosphamide and a human B cell-specific immunotoxin containing pokeweed antiviral protein. Cancer Res. *45:* 69 (1985).

710 Uckun, F.M.; Ramakrishnan, S.; Houston, L.L.: Immunotoxin-mediated elimination of clonogenic tumor cells in the presence of human bone marrow. J. Immunol. *134:* 2010 (1985).

711 Uhlenbruck, G.; Hanisch, F.-G.; Dienst, C.: Die Immunchemie der neuen Kohlenhydrat-Tumormarker; in Wüst, Tumormarker, p. 8 (Steinkopff, Darmstadt 1987).

712 Varki, N.M.; Reisfeld, R.A.; Walker, L.E.: Antibodies associated with a human lung adenocarcinoma defined by monoclonal antibodies. Cancer Res. *44:* 681 (1984).

713 Vaughan, A.T.M.; Bradwell, A.R.; Dykes, P.W.; Anderson, P.: Illusions of tumour killing using radiolabelled antibodies. Lancet *i:* 1492 (1986).

714 Vaughan, A.T.M.; Anderson, P.; Dykes, P.W.; Chapman, C.E.; Bradwell, A.R.: Limitations to the killing of tumours using radiolabelled antibodies. Br. J. Radiol. *60:* 567 1987).

715 Vitetta, E.S.; Cushley, W.; Uhr, J.W.: Synergy of ricin A chain-containing immunotoxins and ricin B chain-containing immunotoxins in in vitro killing of neoplastic human B cells. Proc. natn. Acad. Sci. USA *80:* 6332 (1983).

716 Vitetta, E.S.; Fulton, R.J.; Uhr, J.W.: Cytotoxicity of a cell-reactive immunotoxin containing ricin A chain is potentiated by an anti-immunotoxin containing ricin B chain. J. exp. Med. *160:* 341 (1984).

717 Vitetta, E.S.; Uhr, J.W.: Immunotoxins. Ann. Rev. Immunol. *3:* 197 (1985).

718 Vollerthun, R.; Sedlacek, H.H.; Ronneberger, H.: Gewebeverteilung von nativem und enzymbehandeltem Human-Immunglobulin. Dt. med. Wschr. *102:* 684 (1977).

719 Wagener, C.; Petzold, P.; Kohler, W.; Totovic, V.: Binding of five monoclonal anti-CEA antibodies with different epitope specificities to various carcinoma tissues. Int. J. Cancer *33:* 469 (1984).

720 Wahl, R.L.; Parker, C.W.; Philpott, G.: Improved radio-imaging and tumour localisation with monoclonal F(ab')$_2$. J. nucl. Med. *24:* 316 (1983).

721 Waldmann, H.; Polliak, A.; Hale, G.; Or, R.; Cividalli, G.; Weiss, L.; Weshler, Z.; Samuel, S.; Manor, D.; Brautbar, C.: Elimination of graft-versus-host disease by in vitro depletion of alloreactive lymphocytes with monoclonal rat anti-human lymphocyte (CAMPATH I). Lancet *I:* 483 (1984).

722 Walker, W.H.C.: The Scatchard plot in immunometric assay. Clin. Chem. *23:* 588 (1977).

723 Wang, T.S.T.; Ng, A.K.; Alsedairy, S.; Fawwaz, R.A.; Hardy, M.A.; Alderson, P.O.: A method for direct quantification of the amount of DTPA in ^{111}In monoclonal antibody preparation. Int. J. Radiat. Appl. Instrum., Part A Appl. Radiat. Isot. *38:* 315 (1987).

724 Ward, M.C.; Roberts, K.R.; Westwood, J.H.; Coombes, R.C.C.; McCready, V.R.: The effect of chelating agents on the distribution of monoclonal antibodies in mice. J. nucl. Med. 27: 1746 (1986).

725 Warenius, H.M.; Taylor, J.W.; Durack, B.R.; Cross, P.A.: The production of human hybridomas from patients with malignant melanoma. The effect of pre-stimulation of lymphocytes with pokeweed mitogen. Eur. J. Cancer clin. Oncol. 19: 347 (1983).

726 Watanabe, K.; Matsubara, T.; Hakomori, S.: α-L-Fucopyranosylceramide, a novel glycolipid accumulated in some of the human colon tumors. J. biol. Chem. 251: 2385 (1976).

727 Watanabe, J.-I.; Okabe, T.; Fujisawa, M.; Takaku, F.; Hirohashi, S.; Shimosato, Y.: Monoclonal antibody that distinguishes small-cell lung cancer from non-small-cell lung cancer. Cancer Res. 47: 826 (1987).

728 Watson, D.B.; Burns, G.F.; Mackay, I.R.: In vitro growth of B lymphocytes infiltrating human melanoma tissue by transformation with EBV: Evidence for secretion of anti-melanoma antibodies by some transformed cells. J. Immunol. 130: 2442 (1983).

729 Webb, K.S.; Ware, J.L.; Parks, S.F.; Briter, W.H.; Paulson, D.F.: Monoclonal antibodies to different epitopes of a prostate tumor associated antigen: implications for immunotherapy. Cancer Immunol. Immunother. 14: 155 (1983).

730 Weiner, L.M.; Steplewski, Z.; Koprowski, H.; Sears, H.F.; Litwin, S.; Comis, R.L.: Biologic effects of gamma interferon pretreatment followed by monoclonal antibody 17-1A administration in patients with gastrointestinal carcinoma. Hybridoma suppl. 1, p. 5: 65 (1986).

731 Weinstein, J.N.; Steller, M.A.; Keenan, A.M.; Covell, D.G.; Key, M.E.; Sieber, S.M.; Oldham, R.K.; Hwang, K.M.; Parker, R.J.: Monoclonal antibodies in the lymphatics: Selective delivery to lymph node metastases of a solid tumor. Science 222: 423 (1983).

732 Weiss, R.A.: Retroviruses produced by hybridomas. New Engl. J. Med. 307: 1587 (1982).

733 Welch, M.J.; Welch, T.J.: Solution chemistry of carrier-free indium; in Subramanian, Rhodes, Cooper et al., Radiopharmaceuticals, p. 73 (The Society of Nuclear Medicine, New York (1975).

734 Westerwoudt, R.J.: Improved fusion methods. IV. Technical aspects. J. Immunol. Methods 77: 181 (1985).

735 Wieland, T.; Fahrmeier, A.: Oxydation und Reduktion an der γ-δ-Dihydroxyisoleucin-Seitenkette des O-Methyl-α-Amanitins. Methyl-Aldoamanitin, ein ungiftiges Abbauprodukt. Liebigs Annln Chem. 736: 95 (1970).

736 Wikstrand, C.J.; McLendon, R.E.; Bullard, D.E.; Fredman, P.; Svennerholm, L.; Bigner, D.D.: Production and characterization of two human glioma xenograft-localizing monoclonal antibodies. Cancer Res. 46: 5933 (1986).

737 Williams, M.R.; Perkins, A.C.; Campbell, F.C.; Pimm, M.V.; Hardy, J.G.; Wastie, M.L.; Blamey, R.W.; Baldwin, R.W.: The use of monoclonal antibody 791 T/36 in the immunoscintigraphy of primary and metastatic carcinoma of the breast. Clin. Oncol. 10: 375 (1984).

738 Wood, W.G.; Wachter, C.; Scriba, P.C.: Experiences using Chloramine-T and 1, 3, 4, 6-Tetrachloro-3-6-diphenylglycoluril (Iodogen®) for the radioiodination of materials for radioimmunoassays. J. Clin. Chem. Clin. Biochem. 19: 1051 (1981).

739 Woodruff, M.F.A.: Cellular heterogeneity in tumours. Br. J. Cancer 47: 589 (1983).

740 Wunderlich, D.; Teramoto, Y.A.; Alford, C.; Schlom, J.: The use of lymphocytes from axillary lymph nodes of mastectomy patients to generate human monoclonal antibodies. Eur. J. clin. Oncol. 17: 719 (1981).

741 Wyke, J.A.; Weiss, R.A.: The contribution of tumour viruses to human and experimental oncology. Cancer Surveys *3:* 1 (1984).
742 Yamaizumi, M.; Mekada, E.; Uchida, T.; Okada, Y.: One molecule of diphtheria toxin fragment A introduced into a cell can kill the cell. Cell *15:* 145 (1978).
743 Yeh, M.-Y.; Hellström, I.; Hellström, K.E.: Clonal variation in expression of a human melanoma antigen defined by a monoclonal antibody. J. Immunol. *126:* 1312 (1981).
744 Yeoman, L.C.; Taylor, C.W.; Charabarty, S.: Human colon tumor antigens. Methods Cancer Res. *19:* 234 (1982).
745 Yoda, Y.; Ishibashi, T.; Makita, A.: Isolation, characterization and biosynthesis of Forssman antigen in human lung and lung carcinoma. J. Biochem. *88:* 1887 (1980).
746 Yokota, M.; Warner, G.; Hakomori, S.: Blood group A-like glycolipid and a novel Forssman antigen in the hepatocarcinoma of a blood group 0 individual. Cancer Res. *41:* 4185 (1981).
747 Yoshida, M.; Miyoshi, I.; Hinuma, Y.: Isolation and characterization of retrovirus (ATLV) from cell lines of adult T-cell leukemia and its implication in the disease. Proc. natn. Acad. Sci. USA *79:* 2031 (1982).
748 Yoshitake, S.; Yamada, Y.; Ishikawa, E.; Masseyeff, R.: Conjugation of glucose oxidase from Aspergillus niger and rabbit antibodies using N-hydroxysuccinimide ester of N-(4-carboxycyclohexylmethyl)-maleimide. Eur. J. Biochem. *101:* 395 (1979).
749 Youle, R.J.; Neville, D.M.: Anti-Thy 1.2 monoclonal antibody linked to ricin is a potent cell-type-specific toxin. Proc. natn. Acad. Sci. USA *77:* 5483 (1980).
750 Youle, R.J.; Murray, G.J.; Neville, D.M.: Studies on the galactose-binding site of ricin and the hybrid toxin man6P-ricin. Cell *23:* 551 (1981).
751 Youle, R.J.; Neville, D.M.: Kinetics of protein synthesis inactivation by ricin-anti-thy 1.1 monoclonal antibody hybrids. J. biol. Chem. *257:* 1598 (1982).
752 Yuan, S.-Z.; Ho, J.J.L.; Yuan, M.; Kim, Y.S.: Human pancreatic cancer-associated antigens detected by murine monoclonal antibodies. Cancer Res. *45:* 6179 (1985).
753 Zalcberg, J.R.; Thompson, C.H.; Lichtenstein, M.; McKenzie, I.F.C.: Tumor immunotherapy in the mouse with the use of ^{131}I-labeled monoclonal antibodies. J. natn. Acad. Sci. *72:* 697 (1984).
754 Zimmer, A.M.; Rosen, S.T.; Spies, S.M.; Polovina, M.R.; Minna, J.D.; Spies, W.C.; Silverstein, E.A.: Radioimmunoimaging of human small cell lung carcinoma with I-131 tumor specific monoclonal antibody. Hybridoma *4:* 1 (1985).